MEDIEVAL AND CLASSICAL TRADITIONS AND THE RENAISSANCE OF PHYSICO-MATHEMATICAL SCIENCES IN THE 16th CENTURY

DE DIVERSIS ARTIBUS

COLLECTION DE TRAVAUX
DE L'ACADÉMIE INTERNATIONALE
D'HISTOIRE DES SCIENCES

COLLECTION OF STUDIES
FROM THE INTERNATIONAL ACADEMY
OF THE HISTORY OF SCIENCE

DIRECTION
EDITORS

EMMANUEL
POULLE

ROBERT
HALLEUX

TOME 50 (N.S. 13)

BREPOLS

PROCEEDINGS OF THE XXth INTERNATIONAL CONGRESS
OF HISTORY OF SCIENCE (Liège, 20-26 July 1997)

VOLUME VIII

MEDIEVAL AND CLASSICAL TRADITIONS AND THE RENAISSANCE OF PHYSICO-MATHEMATICAL SCIENCES IN THE 16th CENTURY

Edited by

Pier Daniele NAPOLITANI and Pierre SOUFFRIN

BREPOLS

The XXth International Congress of History of Science was organized by the Belgian National Committee for Logic, History and Philosophy of Science with the support of :

ICSU
Ministère de la Politique scientifique
Académie Royale de Belgique
Koninklijke Academie van België
FNRS
FWO
Communauté française de Belgique
Région Wallonne
Service des Affaires culturelles de la Ville
   de Liège
Service de l'Enseignement de la Ville
   de Liège
Université de Liège
Comité Sluse asbl
Fédération du Tourisme de la Province
   de Liège
Collège Saint-Louis
Institut d'Enseignement supérieur
   "Les Rivageois"

Academic Press
Agora-Béranger
APRIL
Banque Nationale de Belgique
Carlson Wagonlit Travel -
   Incentive Travel House

Chambre de Commerce et d'Industrie
   de la Ville de Liège
Club liégeois des Exportateurs
Cockerill Sambre Group
Crédit Communal
Derouaux Ordina sprl
Disteel Cold s.a.
Etilux s.a.
Fabrimétal Liège - Luxembourg
Generale Bank n.v. -
   Générale de Banque s.a.
Interbrew
L'Espérance Commerciale
Maison de la Métallurgie et de l'Industrie
   de Liège
Office des Produits wallons
Peeters
Peket dè Houyeu
Petrofina
Rescolié
Sabena
SNCB
Société chimique Prayon Rupel
SPE Zone Sud
TEC Liège - Verviers
Vulcain Industries

D/2001/0095/126
ISBN 2-503-51198-8
Printed in the E.U. on acid-free paper

TABLE OF CONTENTS

FOREWORD

Pier Daniele NAPOLITANI - Pierre SOUFFRIN

The texts put together in the present volume were written for the Symposium Medieval and Classical Traditions and the Renaissance of Physico-Mathematical Sciences in the Sixteenth Century which was held in Liège, July 1997, during the XXth International Congress of History of Science : Science, Technology, Industry (IUHPS/DHS).

A glance at the synopsis reveals immediately that of the vast program which the title of the Symposium implied, only two episodes have been touched upon. In no way is this the result of an accidental convergence of the interventions : the first article of the present volume — which one should consider as an introduction to the whole — offers some basic justifications of this fact.

These theoretical reasons have had an organizational and institutional impact on the collaborations, occurring in parallel and often with crossovers, which have involved most of the participants in research programs on the two themes that have been privileged here. These programs of collaborations have occasioned numerous and fruitful exchanges, which intensified themselves, say, in the last decade. The Symposium has been both an episode of collaboration and a convenient opportunity to bring together works that a unity in theme does draw closer. Of course this coherence, that we perceive in these works, should not be confused with any unanimity on analytical methods and on results, which are sufficiently original so that responsibility is to be strictly assumed for each article by its signatory. On this it may not be inappropriate here to say explicitly that the following articles, in most cases, are brief presentations of researches which will be the object of a fuller treatment elsewhere, in professional periodicals and other publications. This is the normal consequence of the fact that each one of the authors of the present volume has strictly observed the space limitations (less than 10 pages/article) imposed by the general editor of this collection of Symposia. May he be here greeted with admiration for his endeavour.

The extension of the historical field implied by the title of the Symposium conveys our conviction, indisputably shared by all participants, that it is indispensable to know better the genuine content of what the naturalist philosophers of the 16th Century knew, if we want to have access to the keys enabling us to understand the emergence of a new science in the following century. Under the circumstances to know better implies, in particular, pondering over the relations between the sciences and the humanities and the historical adequacy of the Medieval traditions based on a meticulous study of the ideas through which and against which were erected the concepts and the theoretical tools that will determine the features of what by convention is called the Scientific Revolution of the 17th Century. We have brought about this Symposium with the conviction that renewed efforts of research concerning original texts are an unavoidable condition to go beyond general discourses constructed on the basis of what may seem philosophically sophisticated but which are historically very tenuous. Such works, — editions, datations and analysis of original texts (so as not to talk of " sources " which would imply a prejudice on the result of the research) — are absolutely required if our discipline, " the history of the sciences ", is to be freed from sterile squabbles between " historians ", " epistemologists ", " sociologists " which today paralyze its development. What is involved here is a particular component of the history of the sciences, and we have the feeling that it is actually illustrated by a group of works, in disciplinary fields especially related to ours. We can mention the colloquium *La nouvelle physique du XIVe siècle...* held in Nice in 1993[1] and the symposia held parallel to ours in Liège in 1997 (to be published by the same editor as the present one) " Ancient, Medieval and premodern Optics " organized by Gérard Simon and A.I. Sabra, and " The East and the West " organized by Roshdi Rashed, André Allard and Chikara Sasaki.

We think — and this binds only the two signatories of the present foreword — that this conception of the history of the sciences must and can prove its requirement as a discipline besides other perfectly legitimate approaches through its production. It is our hope that the works contained in the present volume will contribute to that.

---

1. *La nouvelle physique du XIVe siècle...*, (colloquium held in Nice in 1993), Olschki, 1997.

A L'AUBE DE LA RÉVOLUTION SCIENTIFIQUE : DE GALILÉE À MAUROLICO

Pier Daniele NAPOLITANI

POURQUOI *CE* TITRE POUR *CE* SYMPOSIUM ?

Le lecteur de ce volume se demandera peut-être pour quelle raison ont été réunis ici deux objets d'étude au premier abord aussi éloignés l'un de l'autre, à savoir l'œuvre d'un mathématicien de la Renaissance, Francesco Maurolico, et les réflexions du jeune Galilée sur le mouvement et la philosophie aristotélicienne.

Un premier élément de réponse est assez trivial : plusieurs des personnes impliquées dans l'un de ces deux programmes de recherche — dont nous avons parlé dans l'*Avant-propos* — sont aussi impliquées dans l'autre. Il devenait alors naturel de chercher à organiser une *séance plénière* de ces deux ateliers, et le congrès de Liège offrait une excellente opportunité. Le symposium devait être — et il le fut — le premier temps de convergence de deux expériences jusqu'alors parallèles.

Mais des motifs — comment dire ? — plus *théoriques* nous poussaient à organiser ce moment commun. L'historien éprouve souvent un sentiment d'insatisfaction devant une certaine manière d'affronter la question cruciale de la naissance des sciences physico-mathématiques modernes. Une conviction commune des éditeurs de ce volume — et je crois pouvoir le dire, des auteurs qui ont contribué à son contenu — est l'absolue nécessité d'une reprise des études sur le XVIe siècle. Ces études devraient porter sur les rapports entre science et humanisme, et sur le poids des traditions médiévales ; elles devraient savoir s'immerger dans les textes, dans les problèmes minutieux de datations et d'établissement des sources, et pouvoir fournir de nouveaux instruments de travail à la communauté scientifique. Ces nouvelles études permettraient alors de dépasser les lieux communs, la reprise de thèses vieillies, les études de contexte et les mots d'esprit plus ou moins brillants.

Galilée et son œuvre représentent cependant l'une des origines de ce que l'on a nommé révolution scientifique. Pour dépasser les études antérieures de l'œuvre scientifique de Galilée, il nous semble nécessaire de pouvoir disposer de nouvelles connaissances sur les mathématiques et la philosophie naturelle du XVIᵉ siècle. Pour rester concret et pour des motifs d'espace, nous essaierons d'éclaircir la question à travers un exemple : Galilée et la mathématique archimédienne.

GALILÉE ET ARCHIMÈDE

On a depuis longtemps souligné que l'œuvre du savant pisan faisait de constantes références au " divin ", au " surhumain " Archimède. Mais à quel Archimède Galilée fait-il référence ? Les œuvres du Syracusain furent disponibles pour la plupart à partir de 1544, date de l'*editio princeps* de Bâle, et Galilée travailla et annota cette édition. D'autre part, les premiers pas de Galilée dans la philosophie naturelle et dans la mathématique, semblent s'inspirer de l'édition des *Corps flottants* de Commandin et de son *Liber de centro gravitatis solidorum*. Dans les manuscrits du *De motu antiquiora*, la référence aux *Corps flottants* est constante, et les premiers résultats mathématiques du jeune Galilée furent ceux des *Theoremata de centro gravitatis solidorum* avec lesquels il souhaitait suppléer aux lacunes du livre de Commandin.

La carrière même de Galilée — son poste de lecteur à Pise, sa chaire à Padoue — fut fortement encouragée par l'amitié et l'appui de Guidobaldo Dal Monte. Ce dernier était l'élève et le continuateur de Commandin, et par conséquent un des héritiers et principaux artisans du *revival* archimédien du XVIᵉ siècle. Les rapports entre les deux savants ne se limitèrent pas à un échange de correspondance ou à l'appui apporté par Guidobaldo au jeune Galilée. Les *Meditatiunculae* de Guidobaldo, pour la plupart encore inédites, témoignent d'une profonde communauté de thèmes et d'inspiration.

Tout cela est bien connu et tend à véhiculer une certaine image : Commandin et l'école d'Urbino laissèrent en héritage à Galilée la mathématique d'Archimède, et celui-ci en fit un instrument de construction de sa philosophie naturelle. Mais l'historien qui examine de plus près ces textes se heurte à une série d'insuffisances, d'interrogations, de surprises et de rendez-vous manqués. Les annotations de Galilée à l'*editio princeps* de Bâle ont par exemple été très peu étudiées. Les renvois à Archimède et aux *Corps flottants* que l'on trouve dans la page de la *Bilancetta*, dans le *De motu* et le *Discorso sulle cose che stanno sull'aqua* de 1612, ressemblent plus à des hommages qu'à autre chose. Les démonstrations galiléennes sur les corps flottants prendront toujours des chemins très différents de ceux suivis par Archimède. Elles tourmenteront leur auteur durant de longues années, et se combineront une production de concepts nouveaux et renouvelés — moment, poids spécifique, vitesse, force — et en la modélisation mathématique d'autres phénomènes physiques, parmi lesquels l'équilibre de la balance ou l'étude du mouvement uniforme.

Mais dans tous ces travaux, quelle part prirent et pesèrent les autres traditions ? Peut-on vraiment se borner à considérer l'Archimède de l'école d'Urbino comme l'unique précédent de la mathématique de Galilée ? S'ajoute à cela le fait que les relations entre Galilée et Guidobaldo sont encore en grande partie à étudier. Qui est l'élève, qui est le maître entre le jeune mathématicien et le Marchese Del Monte ? Même s'il nous est impossible de rentrer ici dans les détails, de nombreux passages des *Meditatiunculae* et de la correspondance entre les deux savants — qui mériterait un examen attentif — témoignent d'échanges scientifiques et d'un *parrainage* particulièrement complexes, dans lesquels on entrevoit pour ce qui concerne la formation du jeune Galilée, des interactions de traditions et de facteurs culturels, quand bien même on se restreindrait à la seule formation mathématique.

Plus de soixante ans ont passé depuis qu'Alexandre Koyré affirmait dans ses *Études galiléennes* que " la physique classique, sortie de la pensée de Bruno, de Galilée, de Descartes " aurait eu comme " précurseur et maître ... Archimède " [1] et " l'œuvre scientifique du XVIe siècle pourrait se résumer à la réception et compréhension graduelles de l'œuvre d'Archimède " [2].

Et si on peut partager l'idée principale de cette thèse, celle-ci reste encore — après plus d'un demi-siècle — plus un programme de recherche qu'un résultat historiographique acquis.

QUOT CAPITA, TOT ARCHIMEDES

N'oublions pas que l'œuvre d'Archimède est particulièrement complexe. Elle ne se présente pas comme des " Eléments " de géométrie de mesure, de mécanique ou d'hydrostatique. Elle revêt par contre l'habit d'une série de traités particuliers, écrits dans un style excessivement synthétique, se renvoyant souvent l'un à l'autre, quand ils ne renvoient pas directement à une œuvre introuvable aujourd'hui dans le *corpus* archimédien : œuvres sur les coniques, sur l'équilibre, sur les centres de gravité des solides. Il ne faut pas négliger non plus que l'Archimède d'aujourd'hui est bien différent de celui qu'avait à sa disposition le mathématicien du XVIe siècle. L'Archimède de la Renaissance devait s'appuyer, en dernière analyse, soit sur des copies de la traduction médiévale de Moerbeke, soit sur des copies ou traductions du codex A.

Et si on passa au cours du siècle, des *excerpta* de Valla et de Gaurico à la brillante reconstruction du texte des *Corps flottants* de Commandin, cela n'en fut pas moins un passage difficile, loin d'être achevé quand Galilée commençat ses études mathématiques. Guidobaldo en 1588 ressentait encore la nécessité

1. A. Koyré, *Études galiléennes*, Paris, 1966, 15-16.
2. A. Koyré, *Études galiléennes*, *op. cit.*, 16.

de publier sa propre traduction commentée de l'*Équilibre des figures planes*, et seule une partie du *corpus* archimédien était disponible avec autant de clarté et de précision que dans les traductions de Commandin. En résumé : l'assimilation de l'héritage archimédien ne se fit pas instantanément ; ce fut plutôt un long processus auquel participèrent plusieurs mathématiciens et humanistes : Valla, Gaurico, Maurolico, Tartaglia, Venatorius, Commandin, Guidobaldo. Chacun d'eux — et d'autres encore —, confronté à l'œuvre d'Archimède, fut porté, obligé même, à produire son propre Archimède, sa propre interprétation.

Si l'on sort du domaine strictement mathématique ou philologique, les choses deviennent encore plus complexes. En particulier pour ce qui concerne les corps flottants, les centres de gravité et la mécanique. Les thématiques les plus spécifiquement archimédiennes se télescopaient alors avec les traditions scolastiques, la science des poids et les *Quaestiones mechanicae* aristotéliciennes à peine redécouvertes. À l'intérêt d'Archimède pour le mathématicien ou l'humaniste, s'ajoutait celui du technicien, du militaire, du philosophe. On pensera bien sûr à Tartaglia et à ses éditions, mais aussi à un des maîtres pisans de Galilée, Francesco Bonamici, qui — ainsi que l'a montré Mario Helbing — cherchait à utiliser les descriptions archimédiennes du phénomène de la flottaison dans un cadre explicatif et causal aristotélicien.

De Maurolico à Galilée

Le lecteur aura maintenant compris où nous voulions en venir avec cet exemple : l'association des figures de Maurolico et Galilée dans un même symposium, dans un même volume, loin d'être seulement due à une simple opportunité d'organisation, voulait, et veut être encore une proposition pour relancer les études sur le XVIe siècle *dans son ensemble*. Et de ce point de vue, Maurolico représente un point de passage obligé.

Le mathématicien de Messine ne fut pas seulement l'un des artisans de la mathématique de redécouverte des Anciens. Son attitude fut très différente de celle de son contemporain Commandin — ce dernier se concentrant surtout sur le recouvrement philologique des textes anciens —, ou celle d'autres mathématiciens qui furent les acteurs de cette redécouverte. Maurolico fut, à la différence d'eux, un mathématicien doté d'une grande créativité et originalité, n'hésitant pas à intervenir sur les textes qu'il étudiait pour les reconstruire entièrement sur des bases nouvelles.

L'originalité de ses résultats s'est trouvée assombrie par toute une série de facteurs : la publication tardive de nombreuses de ses œuvres, l'apparente conformité au modèle classique et un environnement culturel relativement décentré, Messine. Mais pour rester dans le cadre de l'exemple que nous avons pris jusqu'à maintenant, on n'oubliera pas que ce fut justement Maurolico qui le premier donnat un modèle mathématique pour le concept de moment mécanique. Il fut aussi le premier à tenter une révision de l'œuvre d'Archimède, cherchant à en extraire des méthodes démonstratives nouvelles et plus efficaces. La structure mathématique de sa version de la *Sphère et le Cylindre* est très différente du texte d'Archimède ; son *De momentis aequalibus*, plus qu'une simple édition de l'*Équilibre des figures planes*, doit être considéré comme une œuvre presque entièrement originale ; et il en est de même, certes à des degrés divers, pour nombre de ses travaux.

En résumé : les études archimédiennes de Maurolico représentèrent dans le contexte mathématique des trois premiers quarts du XVIe siècle, une alternative à l'attitude des cercles de Bâle à l'origine de l'*editio princeps* de 1544 — caractérisée surtout, semble-t-il, par la simple récupération de la lettre du texte grec. Ils représentèrent tout autant une alternative aux tentatives de Tartaglia pour retrouver les œuvres d'Archimède dans la culture de l'abaque, ainsi qu'à l'attitude de Commandin, héritier de la tradition humaniste, qui souhaitait leur rendre leur statut de référence absolue et intouchable.

Il est vrai que les efforts de Maurolico dans ce domaine restèrent inédits jusque en 1685. Mais cela ne signifie pas autant qu'ils restèrent lettre morte. Commandin et Maurolico entretinrent une correspondance scientifique, dont seulement un fragment nous est parvenu. Ce fut surtout par l'intermédiaire des Jésuites et particulièrement de Clavius, que ses idées — même si cela se fit peut-être par d'autres truchements que les textes — semblent avoir connu cependant une certaine diffusion dans le dernier quart du XVIe siècle. Leur influence sur Luca Valerio est quasi-certaine. On ne peut non plus exclure que le jeune Galilée ait pu d'une manière ou d'une autre être influencé par les idées de Maurolico dans ses premiers travaux sur les centres de gravité.

***

De ce point de vue, les études sur Maurolico se trouvent liées à notre compréhension de la formation mathématique de Galilée. Entendons nous : nous ne voulons certes pas soutenir la thèse d'une influence directe du mathématicien de Messine sur les premières travaux de Galilée. Même si elle était avérée, elle serait de toute façon plus probablement la conséquence de tout un réseau de circulation de textes et d'idées.

Le point que nous souhaitions illustrer était plutôt l'importance de la reprise des études mathématiques au XVIe siècle. Et l'on concédera alors à Maurolico un rôle de premier plan.

GREEK HERITAGE AND THE SCIENTIFIC WORK
OF FRANCESCO MAUROLICO

Rosario MOSCHEO

Francesco Maurolico (1494-1575), a Sicilian mathematician and humanist, is the main character I am concerned with in my studies on the southern Italian sources of the scientific thought of the sixteenth and seventeenth centuries. Attention to him has grown widely in the last few decades, because of new and important scholarship on particular topics such as the Medieval and the Renaissance tradition of Archimedes, and because of a renewed general interest by historians of science in the sixteenth century, as the great transitional period between the humanistic rediscovery of Greek science and the Galileian revolution[1].

Maurolico's highest credit has been that of having contributed (as much as Federico Commandino) to the vigorous revival of Mathematics, through a careful as well as systematic recovery of what had been already achieved by the Greeks in this field. Such a contribution is fully documented by Maurolico's program of editing ancient mathematical texts and by the parallel production of original treatises of his own. In summary, a great deal of work of Maurolico is still extant ; a glimpse of this can be easily obtained simply by looking at the *Index lucubrationum Maurolyci*, a lengthy list whose many versions, quite often due to Maurolico himself, represent more than half a century of continuous effort in science, which he carried out exclusively in Sicily[2].

The aim of these pages is to give some information on the relationship between Maurolico's work and his own Sicilian environment from the standpoint of the Greek scientific learning. I propose myself to explore this relationship to some extent, and to find out in what ways the Sicilian cultural enclave

---

1. R. Moscheo, *F. Maurolico tra Rinascimento e scienza galileiana. Materiali e ricerche*, Messina, Società Messinese di Storia Patria, 1988 (= Biblioteca dell'Archivio Storico Messinese, X).

2. R. Moscheo, *Mecenatismo e scienza nella Sicilia del '500 : F. Maurolico ed i Ventimiglia di Geraci*, Messina, Società Messinese di Storia Patria, 1990 (= Biblioteca dell'Archivio Storico Messinese, XIV).

could have influenced or also made possible Maurolico's production. The true interest of questions like these lies mainly in the two following points :

1) the largely marginal position of Sicily in the general cultural context of the sixteenth century ;

2) the remarkable amount of activity of Maurolico, who carried out his large program of scientific renewal by staying in his island and, therefore, by remaining very much isolated in his work, and far away from the possibility of having much more productive contacts with other important and well established scientific environments, such as those of Venice and Tuscany.

It is fairly clear that both these points represent the opposite as well as contrasting sides of the same medal. Nevertheless this contrast is somehow an apparent one. In fact, the lack of a proper environment is regrettably striking in the whole of Maurolico's intellectual biography. For instance, a patronage similar to that enjoyed in other areas by Commandino, would have allowed Maurolico to publish regularly in his life time his own *lucubrationes*, so enhancing further his influence on the contemporary science.

The roots of Maurolico's work are largely to be traced in Sicily, and specifically in that particular atmosphere which was characteristic of Messina at that time. This included certain well known Greek institutions, such as the basilian monasteries, and the parallel existence of a famous school which, partially attached to those institutions, reached its climax towards the end of the preceding century, under the rule of a well known Byzantine refugee : Constantine Lascaris.

As a matter of fact, the existence in Messina of such a school[3], the aims of which were not simply that of teaching the Greek language to unwilling monks, but also — especially at Lascaris time — the spreading of knowledge to a wider audience, of both Greek literature and philosophy, " explains " thoroughly what can be defined as the phenomenon named " Maurolico " ; that is the rise of a scholar who seems to have been in many respects the most important outcome, although late, of that institution.

In what follows, leaving aside Maurolico's biography, I will mainly try to identify some of the " scientific connections " in Lascaris teaching as well as in his much better known career as a collector of Greek manuscripts.

THE RENAISSANCE SCIENCE AND CONSTANTINE LASCARIS

According to a rather obscure legend, Lascaris's death — the date of which is here advanced by quite a few years — came as a consequence after a great excitement he went through on an occasion, when, while he was explaining to

3. R. Moscheo, " L'insegnamento del greco a Messina ' dopo ' C. Lascaris (note in margine ad una pergamena inedita) ", *Nuovi Annali della Facoltà di Magistero dell'Università di Messina*, 5 (1987), 537-550.

his pupils, from Plato' *Timaeus*, the famous passage relating to Atlantis, a breathless messenger, suddenly entering the classroom, brought the news concerning the successful outcome of Columbus enterprise. In Lascaris mind, the unexpected as well as untaught of possibility of seeing somehow verified in the real world, and in such an astonishing manner, what was for him uniquely a charming myth, must have caused — I think — together with the excitement, the sudden rise of the conscience of a new highly important dimension in his own teaching. A " practical " dimension which has previously been only latent just because, according to a quite general habit, the deep enthusiasm by Lascaris towards the classical word has been too much affected by the empty rhetoric of most of the humanist tradition.

Although the historical truth is completely different, the image of Lascaris which comes out of the legend is truly a real one. Some well sound biographical data denies the whole story : for instance, the death of Lascaris sometime around the 1493 and its immediate although indirect cause. However, what it is here really interesting is that, for the first time, and in a peculiar way, the portrait of Lascaris that one gets is absolutely different from the stereotype which has been used so far by the historians. Namely, the stereotype of a well known teacher of Greek, author of an even better known Greek grammar. The hero of the legend, not indeed the grammarian, is in fact a learned man, a teacher of philosophy, an extremely sensitive person not only to the very many impulses which derived from his own erudite background, or from the deep familiarity he had with the literary as well as philosophical " monuments " of the Greek and Byzantine traditions to which he felt himself totally belonging. He was also a man who was open minded, sensitive to the experiences of his own contemporary world, to its political struggles, to the theological and philosophical novelties, to the great geographical changes and therefore to the scientific progress of his age.

But, what can we really say about the genuine scientific interests of Constantine Lascaris ? Their identification is quite difficult because of the lack of any detailed study on his literary relationships and on his library. In fact, I must correct this last statement : if the most recent attempt I know of in reconstructing Lascaris manuscript collection, is a well known paper published in 1966 by José Maria Fernández Pomar[4], an entirely new approach is now represented by the monograph on *Konstantinos Laskaris Humanist, Philologe, Lehrer, Kopist*, published in 1994 by Teresa Martínez Manzano[5].

Mainly because of its prevalent paleographical approach the essay by Fernández Pomar had left substantially unsatisfied the need for a reliable map

4. J.M. Fernández Pomar, " La colección de Uceda y los manuscritos griegos de C. Láscaris ", *Emerita*, 34 (1966), 211-288.

5. T. Martínez Manzano, *Konstantinos Laskaris Humanist, Philologe, Lehrer, Kopist*, Hamburg, 1994 (= *Meletemata. Beiträge zur Byzantinistik und Neugriechischen Philologie*, Band 4, hrs. von A. Kambylis).

and an exact chronology of all the cultural interests Lascaris had throughout his long career. Not a single mathematical paper, nor any specific scientific research of his own emerged from the bulk of data which have been published by Fernández Pomar. And most of the works of Lascaris turns out to be exclusively linked to Greek literature ant to his employment as a teacher of Greek grammar.

Nevertheless, as Martínez Manzano and myself have indipendently proved, the same bulk of data contains other useful informations which allow us to find evidence of some true scientific interests in Lascaris's life. That is what I propose to do in three different stages : a) by looking at his library ; b) by analysing his surviving correspondence, and c) by using some other data which come out of some contemporary or slightly later sources.

a) Lascaris's library

Apart from his well known Greek grammar, the first entirely Greek book which has been printed (and it may be interesting to know that this grammar is the book which Raphael Itloideus brings to the inhabitants of Utopia in the celebrated essay by Thomas More), the collection of Greek manuscripts is the most important surviving relic of Constantine Lascaris. The collection was presented by him to the city of Messina (very likely following the example of Bessarion's similar bequest to the Republic of Venice), where it remained as long as two centuries more or less. In 1678, after the failure of a long revolt of the town against the central Government, the collection was confiscated by the Spaniards and brought a few years later to Madrid. In 1712, Lascaris books eventually entered the Madrid Royal Library, at present the Biblioteca Nacional. There is no information about the actual size of the collection at the time of Lascaris' death (1501). Indeed, nothing is known about it until the collection's entry in the Royal Library. Nevertheless, the scattered existence of quite a few other manuscripts by Lascaris in various modern European libraries, makes it certain that Lascaris's collection was originally much larger than it appears at present.

A rapid survey of all the information which is contained in the Juan Iriarte's Catalogue of the Greek manuscripts held in Madrid's Royal Library, in the papers by Fernández Pomar, as well as in another paper on Lascaris's collection by Gregorio de Andrés[6], is good enough to make us sure that, as much as in any humanistic library of the fifteenth century, Lascaris's collection was not lacking at all in scientific manuscripts. Astronomical as well as astrological and hermetic treatises ; works on natural science, on medicine ; fragments of theoretical works on music, as well as arithmetical treatises, are well represented in this collection. This simple evidence (more details on Lascaris's

---

6. G. de Andrés, " Catálogo de los códices manuscritos de la Biblioteca del Duque de Uceda ", *Revista de Archivos, Bibliotecas y Museos*, LXXVIII (1975), 5-40.

manuscripts can be given at the end of my talk) suggests a tentative evaluation of Lascaris collection in terms of the different disciplines which are there represented.

In my attempt I will focus attention on the 92 madrilenian Lascaris manuscripts. Seventy five of these (the 80%) are individually homogeneous with respect to their content. The remainder can be split into thirty-seven homogeneous parts. Therefore, the brief analysis I am going to present is based on the total number of 112 homogenous " units ". I add here in advance that the extension of the same analysis to include the other known not madrilenian manuscripts of Constantine Lascaris, does not affect sensibly the results here given. Furthermore, it is now practically impossible to draw precise distinctions between what a fifteenth century scholar would have regarded as science or philosophy. Indeed, the official date of birth of both modern science and philosophy is usually fixed at the time when the distinctions became really possible. Therefore, I will deal jointly in my analysis with the scientific and philosophical items.

With the conventions established above, the madrilenian manuscript " corpus " of Constantine Lascaris can be classified in five groups according to the following scheme :

Groups	Units	Percentage
1. Literary writings	46	41
2. Philosophical and scientific writings	43	38
3. Grammatical writings	10	9
4. Theological writings	7	6
5. et al.	6	5
	112	100

This identity card of Lascaris's collection, although approximate, is quite important. While it confirms its predominantly literary character, it suggests also an almost equal importance of the philosophical and scientific section. The grammatical section is far behind the first two. Furthermore, its real weight with respect to the general composition of Lascaris's library turns out to be weaker than that indicated by the figures given above. This is because, the teaching of the Greek, as an official duty of Constantine (to be precise, the duty on which his own life support mainly depended), may certainly reflect, if not his largely fictitious grammatical interests, a related inflated number of grammatical writings in his collection : 9%, even though it is a small figure may in fact give an inflated impression, since grammar was his professional work rather than a great personal interest.

Let me now look more closely, for a while, at the philosophical and scientific section. The simple mention of all the authors whose writings are there represented does not give a full idea of its real content. We have in fact, as much as in the other sections of Lascaris's library, a great amount of excerpts of various kind. These excerpts are quite often in Lascaris's hand, and taken by him in some cases from unidentified writings or unidentified authors. This makes it very difficult to achieve an accurate description of the whole section. Apart from this substantial portion, I can indicate the presence in such a section of authors like Adamantius, Aratus, Diophantus, Dioscorides, Eratosthenes, Galen, Geminus, Hipparchus, Hippocrates, Nicomachus of Gerasa, Oppianus, Simeon Seth, Synesius. The works by Plato and Aristotle are of course well represented in this section, together with an extended list of commentaries, ranging from the classic Age to the medieval Byzantium.

The inner chronology of this section, on a palaeographical base, shows a continuous growth in time of the Lascaris related cultural interests. Such a growth, far from being haphazard, is meaningful in itself. Indeed, it shows the direction of change of these interests, from a preference (apparent, at least) for Aristotle to a preference for Plato. The first preference being typical of Lascaris old age, and specifically of the time of his stay in Messina. Although this evidence needs to be properly understood, I am not giving here any interpretation of it. However, any attempt in this direction should be based on what follows. If, indeed, the Aristotelian writings seem to interest much more the young Lascaris, this is true only, at least during his stay in Milan, with regard to the ethical works. As a matter of fact, Lascaris continued in Messina his search for Aristotelian texts, but at this time his attention was uniquely directed to the scientific works of Aristotle, and the related commentaries.

b) From Lascaris's correspondence

What has been said so far seems sufficient to give at least a rough idea concerning the actual breadth of Lascaris own cultural interests. This is achieved through a careful identification of each individual text, and of his probable readings. By the way, it is also true that exactly the same approach shows unambiguously the significant " blanks " in his library. The philosophical and scientific section, mainly in the latter part, does not represent an exception in this connection. Indeed, it has its own dreadful " blanks ".

Although mathematics is present in Lascaris' collection with authors such as Nicomachus and Diophantus (both of them in an important manuscript of the fourteenth century, the Matr. 4678), there is no doubt that other mathematicians, such as Archimedes, Apollonius, Euclid, Eutocius, Pappus, Serenus, etc., are completely absent. Furthermore, although astronomy is represented through the treatises by Aratus, Geminus, and other brief anonymous tracts, full of scholia, and the poem by Aratus is accompanied by the important scientific comment by Eratosthenes, it is undoubtfully true that astronomers such as Ptolemy, Theodosius, Menelaus, Aristarchus, etc. are lacking in this collection.

In summary, Lascaris' library, although containing a scientific sector, was completely lacking in a series of authors who were very important indeed for the "next coming" scientific revolution, authors who were strongly represented in many other contemporary libraries such as those belonging to other Byzantine scholars, as well as others belonging to various exponents of Latin humanism.

Lascaris himself was perfectly aware of the gravity of such gaps in his library, and he did not miss opportunities during his life time to fill them up. As far as the philosophical and scientific section is concerned, his known attempts in this direction date to his last years. In a letter to Jacopo Antiquario — a letter which, according to Heiberg, has been written in 1492 — Giorgio Valla, a former pupil of Lascaris during his stay in Milan, writes that having been requested by his former teacher (*olim praeceptor meus*) for lists of his books, especially of the mathematical ones, he decided not only to prepare these lists, but to get also copies of the most important manuscripts in his collection, and finally to send these copies to Sicily.

There is no hint which indicates that Valla's intentions were realised. However, the presence in the Madrid corpus of Lascaris books, of a manuscript of Hippocrates (now the Matr. 4634, with the *Aphorisms* and the *Prognosticha*), which had been written by Valla towards the end of the fifteenth century (Fernández Pomar is not clear on this particular point), makes it possible that Valla may have answered positively to Lascaris' request, and that he sent the manuscript of Hippocrates, not only to help the good health of his master (not *ad sanitatem eiusdem tuendam*), but specifically as among the "some books" (*aliquot volumina*) which he had mentioned, without any further indication, in his letter to Antiquario. The letter by Lascaris which Valla is referring to is not extant. The surviving correspondence between them, edited in Migne's *Patrologia Graeca* vol. 161 and integrated by Heiberg, does not show any significant reference to books which may be interesting in our contest.

c) *Other sources*

The existence alone of large gaps in Lascaris's library (I refer mainly to its scientific section) can not explain easily what, in the quoted letter by Valla, seems the sudden wish in Lascaris mind, now that of an old man, and not in a good health, to fill them up. All this can have, perhaps, just one explanation. The revival of those cultural interests which the master had been forced to suppress for so long because of his strong commitments for a long time to teaching Greek grammar to unwilling monks who delayed payment chronically. A revival which seems very likely due to the appearance of a group of new students external to the monastic environment. because of this, from 1492 onwards, the school of Lascaris became comparable to a centre for advanced studies. I am referring here to a small group of venetian, the most prominent of which was the future cardinal Pietro Bembo.

This group went to Sicily because of the good reputation of Lascaris, and above all because of the previous relationships which Lascaris had already established with the venetian and lombard humanism, through his earlier milanese students. Their presence made a substantial contribution towards consolidating those relationships. See, for instance, the subsequent event of the Aldine edition of Lascaris's Greek grammar. All of this allowed Lascaris school to overcome the narrow statutory limits which confined it to the glorious, but now also suffocating guardianship of the great Monastery of San Salvatore[7].

The new phase in the life of the school, mainly because of the enthusiasm and the fresh energies of the new pupils, was probably expressed in a differentiation and in a renewal of his teachings. Now science and philosophy seem to be taught. And, if Lascaris, or the monasteries or other institutions were lacking in suitable books, these could well have been supplied by the most prominent pupils.

Mathematics and astronomy, in particular, may have been taught by Lascaris at this time, as well as the Greek language. This is explicitly suggested by a piece of information which concerns Antonio Maurolico, the father of Francesco. Bembo himself may have studied some mathematics under Lascaris, and this is suggested by his own juvenile scientific interests which include studies on Euclid, and on natural science. (Bembo's juvenile dialogue *De Aetna*, apart from its well known aesthetic values, has its own philological and scientific basis in the *Castigationes plinianae* of Hermolao Barbaro, another venetian friend of Lascaris). With regard to Bembo, another interesting particular, concerning his relationship with Lascaris is given by Maurolico in the first Dialogue of his *Cosmographia*[8]. He describes there an armillary sphere made of bronze, having a bilingual indication (Greek and Latin) of the signs of the zodiac. This instrument had been sent by Bembo, on his way back to Venice (in 1494) to Lascaris at Messina, together with other unspecified gifts. One may guess (wrongly) that astronomy may not have been a real interest of Pietro Bembo, but there is no doubt that his gift of a sphere represents a real interest of Constantine Lascaris in that field[9].

7. M.B. Foti, *Il monastero del S.mo Salvatore in lingua Phari. Proposte scrittorie e coscienza culturale*, Messina, 1988.

8. F. Maurolico, *Cosmographia*, Venice, 1543.

9. R. Moscheo, " Scienza e cultura a Messina tra '400 e '500 : eredità del Lascaris e " filologia " mauroliciana ", *Nuovi Annali della Facoltà di Magistero dell'Università di Messina*, 6 (1988), 595-632.

LA FORMATION DU JEUNE MAUROLICO ET LES AUTEURS CLASSIQUES

Roberta TASSORA

LES PREMIERS PROJETS EDITORIAUX DE FRANCESCO MAUROLICO

L'oeuvre scientifique de Francesco Maurolico[1] (Messine 1494-1575) a été caractérisée par l'évolution continue d'un projet d'élaboration et de systématisation du savoir mathématique. Dans l'intention de l'auteur ce projet aurait dû se développer à travers la redécouverte des oeuvres les plus importantes écrites par les grands mathématiciens de l'Antiquité, dont les contenus auraient dû se fondre avec les nouvelles connaissances des chercheurs modernes. De ce point de vue Maurolico s'insère dans l'esprit culturel de son époque, mais, en même temps, il s'en éloigne par la volonté de redécouvrir les anciens à travers les idées nouvelles du mathématicien moderne. Il ne s'agit donc pas d'une reconstruction philologique, mais plutôt d'une oeuvre tout à fait mathématique que Maurolico illustre dans les différentes versions de son projet.

La conception, et plus encore, la réalisation de ses idées éditoriales supposaient une bonne connaissance mathématique de la part de Maurolico et une large disponibilité des matériaux d'étude difficile à comprendre si l'on pense à la position marginale de Maurolico par rapport aux centres culturels les plus

---

1. Pour une biographie de F. Maurolico on peut voir : *Vita dell'Abbate del Parto D. Francesco Maurolyco scritta dal Baron della Foresta, ad istanza dell'Abbate di Roccamatore D. Silvestro Marulì fratelli di lui nipoti*, Messina per P. Brea, 1613 ; G. Macrì, *F. Maurolico nella vita e gli scritti*, Messina, 1901 ; M. Clagett, *Archimedes in the Middle Ages*, Madison, Philadelphia, 1964-1984 ; Pour une chronologie et une analyse des oeuvres mauroliciennes : M. Clagett, " The Works of F. Maurolico ", *Physis*, XVI (1974), 150-198 ; R. Moscheo, *F. Maurolico tra Rinascimento e Scienza galileiana. Materiali e ricerche*, Messina, 1988 ; F. Napoli, " Intorno alla vita e ai lavori di Francesco Maurolico con appendice di scritti inediti ", *Bullettino di bibliografia e di storia delle Scienze matematiche e fisiche*, XI (1876), 1-121 ; Pour l'analyse de la figure de Maurolico à l'intérieur de la Renaissance des mathématiques : P.L. Rose, *The Italian Renaissance of Mathematics*, Genève, 1975 ; P.D. Napolitani, *Maurolico e Commandino*, Atti del convegno su *Il meridione e le scienze*, Palermo, 1988, 239-258 ; Pour une bibliographie complète : R. Moscheo, *Francesco Maurolico tra Rinascimento e Scienza galileiana. Materiali e ricerche*, Messina, Società Messinese di Storia Patria, 1988.

importants d'Italie. C'est pour cela que les historiens qui étudient l'oeuvre de Maurolico doivent chercher à résoudre le problème des sources du savoir scientifique que Maurolico montre avoir acquis lorsqu'il écrit son projet de reconstruction.

De ce projet ambitieux Maurolico a donné plusieurs versions différentes qui témoignent de l'évolution de la préparation mathématique de l'auteur et de l'approfondissement de ses connaissances.

En particulier je vais vous parler des premiers projets élaborés par le jeune Maurolico entre la dernière partie des années vingt et la première moitié des années trente. En parlant des projets de sa jeunesse je chercherai à vous exposer mon idée sur le développement de la connaissance scientifique de Maurolico pendant cette première période notamment en ce qui concerne les études mauroliciennes sur les sections coniques.

Au cours de nos études nous avons analysé les deux premières versions du projet mathématique de Maurolico qui nous sont parvenues : la première se trouve dans la préface à un texte de grammaire, intitulé *Grammaticorum rudimentorum libelli sex*[2], publié à Messine en 1528 ; la deuxième est contenue dans une lettre au Cardinal Pietro Bembo de 1536[3]. Dans cette lettre Maurolico décrit le total état d'abandon des études mathématiques et il présente son projet de restauration des textes des anciens visant à redécouvrir les auteurs malheureusement oubliés.

Le programme éditorial de 1528 propose l'élaboration de certains des plus importants textes classiques et l'idée de la compilation finale d'un précis à partir des ouvres d'Euclide, Théodose, Ménélaüs, Archimède, Jordanus, Boèce, Ptolémée et d'autres illustres mathématiciens[4]. Bien sûr, l'ampleur du savoir mathématique et des auteurs que Maurolico paraît connaître déjà en 1528 nous surprend, mais en même temps nous devons remarquer que, parmi les grands mathématiciens de l'Antiquité, Maurolico ne mentionne jamais Apollonius, ou Serenus. Cette observation est très importante de notre point de vue étant donné que Apollonius et, d'une façon bien sûr moindre, Serenus sont les auteurs des oeuvres sur les sections coniques les plus importantes de l'Antiquité. Il faut dire que la seconde version du même projet, celle de 1536, dont nous avons parlé, est encore plus riche et elle prévoit aussi la reconstruction des oeuvres d'Apollonius et de Serenus[5]. En effet la période entre 1528 et 1536 semble avoir été caractérisée par un important développement des connaissances mauroliciennes : pendant cette période, par exemple, Maurolico a

2. *Francisci Maurolyci Presbyteri Messanensis grammaticorum rudimentorum libelli sex*, Messina, 1528, 7-8.

3. Cette lettre a été publiée dans sa version intégrale par R. Moscheo dans son *F. Maurolico tra Rinascimento e Scienza galileiana. Materiali e ricerche*, 271-275.

4. Le texte latin de ce projet se trouve en annexe I à la fin de la communication.

5. Un extrait de la lettre de 1536 au Cardinal P. Bembo se trouve en annexe II à la fin de la communication.

connu aussi un texte sur la quadrature de la parabole et peut-être une oeuvre sur le miroir ardent. En effet dans le premier projet, parmi les oeuvres d'Archimède, Maurolico ne cite que le *De circuli dimensione*, le *De sphaera et cylindro* et le *De momentis aequalibus*, tandis qu'en 1536 il introduit aussi le *De quadratura parabolae*, le *De speculis ignificis*, et le *De isoperimetris*.

Nous pouvons, donc observer que Maurolico a inséré, non seulement les oeuvres d'Apollonius et de Serenus sur le sections coniques, mais aussi des textes d'Archimède ou pseudo-archimediens qui révèlent la connaissance d'une théorie sur les sections du cône, en particulier sur la parabole.

Dans son premier projet Maurolico avait déclaré avoir reconstruit certains traités d'Archimède sans connaître les textes originaux et il faisait référence aux *De circuli dimensione, de sphaera et cylindro* et au *De momentis aequalibus*. La déclaration déconcertante de Maurolico sur la reconstruction de certaines oeuvres sans avoir aucune source nous semble vraiment très singulière et, en effet, les études réalisées à ce sujet ont montré que ce genre d'affirmation doit être interprété dans le sens que Maurolico ne connaissait pas le texte original, mais des versions très altérées. Les études développées jusqu'à présent indiquent que le jeune Maurolico devait travailler avec des matériaux qui ne reproduisaient les textes originaux que d'une façon partielle et quelquefois avec beaucoup de transformations.

La recherche des sources des premières oeuvres mauroliciennes a fait apparaître un groupe d'oeuvres écrites en 1534 qui montrent une dépendance vis à vis de l'oeuvre encyclopédique de Giorgio Valla[6] *De expetendis et fugiendis rebus opus* publiée a Venise en 1501. Il s'agit d'une oeuvre qui pour son caractère encyclopédique a le mérite de mettre à disposition du lecteur un vaste matériau d'étude même s'il est quelquefois difficile de trouver une explication du choix des sujets traités. La connaissance de l'oeuvre de Valla aurait permis à Maurolico de connaître, au moins d'une façon superficielle, plusieurs oeuvres de la mathématique classique. Et, en effet, il est possible d'affirmer que Maurolico connaissait au moins en 1534 le *De expetendis*. Nous avons montré sans aucun doute que le *Sereni cylindricorum libelli duo* a été reconstruit par Maurolico à partir du chapitre du XIIIe livre du *De expetendis* : *de cylindrica sectione*[7]. De l'autre côté Mogenet dans son étude sur l'oeuvre d'Autolycos avait montré une dépendance de l'ouvrage de Valla aussi pour les traités *De sphaera quae movetur* et *De ortu et occasu syderum*[8]. En plus, pour

6. Pour la figure de G. Valla (Piacenza 1447-Venise 1500), humaniste avec des intérêts scientifiques aussi, on peut voir : J.L. Heiberg, " Beiträge zur Geschichte George Valla's und seiner Bibliothek ", *Zentralblatt für Bibliothekswesen*, Beiheft 16 (Leipzig, 1896) ; P.L. Rose, *The Italian Renaissance of Mathematics, op. cit.* ; G. Gardenal, " Giorgio Valla e le scienze esatte ", Etudes de G. Gardenal, P. Landucci Ruffo, C. Vasoli présentées par V. Branca, *Giorgio Valla tra scienza e sapienza,* Firenze, 1981.

7. A ce sujet on peut voir : R. Tassora, " I. *Sereni cylindricorum libelli duo* di Francesco Maurolico e un trattato sconosciuto sulle sezioni coniche ", *Bollettino di Storia delle Scienze Mathematiche*, vol. xv, fasc. 2 (1995), 135-264.

ce qui concerne le *De sphaera et cylindro* on a que la proposition sur le problème de Dionisodoro de couper une sphère selon un rapport donné (prop. XXXII *aliter*) a été prise par Maurolico d'un chapitre du *De expetendis*[9].

Notre étude sur l'emploi des sections coniques dans les *Photismi de lumine et umbra*, un traité d'optique écrit par Maurolico en 1521, a montré qu'il connaissait déjà en 1521 un théorème des *Coniques* d'Apollonius, le théorème sur la section subcontraire du cône, qui n'était présent que dans l'oeuvre de Valla[10].

Il convient donc de remarquer l'importance que la connaissance du *De expetendis* a eue pour le développement des premières connaissances scientifiques de Maurolico et de son programme éditorial. Mais en même temps le caractère non exhaustif de l'oeuvre de Valla, et peut-être des autres sources employées par Maurolico, a déterminé le besoin de combler les lacunes par une recherche mathématique personnelle et par ses propres idées.

LES *ELEMENTA CONICORUM* DU JEUNE MAUROLICO

Ce que je viens de dire d'une façon générale est bien illustré par un cas que j'ai étudié en particulier. Je me réfère à l'attitude de Maurolico face à la nécessité de connaître et employer une théorie sur les sections coniques. En effet, si nous regardons attentivement les oeuvres réalisées par Maurolico en 1534 nous pouvons bien comprendre son besoin de se référer à une théorie quelconque sur les sections du cône. Le traité sur la section du cylindre et celui sur la quadrature de la parabole ne pouvaient pas être pensés et écrits sans la connaissance par leur auteur des propriétés principales des sections coniques.

D'autre part déjà dans les *Photismi* Maurolico employait des propriétés, bien sûr très simples, des sections coniques. On peut donc conclure que pour écrire ses premiers traités Maurolico devait connaître quelque chose sur les sections du cône. Mais alors il faut se poser la question de ce que Maurolico pouvait avoir lu à ce propos. Quel genre de théorie employait-il dans ses ouvrages de jeunesse? Dans le monde occidental la connaissance de l'oeuvre d'Apollonius avait été très superficielle pendant tout le Moyen Age et les savants ne connaissaient que quelques propriétés élémentaires jusqu'à la moitié du XVIe siècle. En effet, la première traduction latine des *Coniques* n'a été

8. J. Mogenet, *Autolycus de Pitane*, Louvain, 1950.

9. M. Clagett, *Archimedes in the Middle Ages, op. cit.*, vol. III, partie III, 1170.

10. Pour un aprofondissement de cette question on peut voir R. Tassora, *Il giovane Maurolico e lo studio delle sezioni coniche*, Tesi di Laurea, Université de Pise, année universitaire 1994-1995, rapporteur Prof. P.D. Napolitani.

publiée qu'en 1537[11]. Mais les ouvrages mauroliciens dont nous sommes en train de parler datent de 1534 !

Les résultats de notre analyse des références aux coniques présentes dans les traités mauroliciens précédant 1534 sont résumés dans le schéma qui suit (p. 30).

J'ai représenté toutes les oeuvres dont je vais parler et j'ai introduit des flèches différentes pour indiquer une relation source-nouvel ouvrage ou bien référence à un texte. Avant de vous expliquer le schéma je voudrais souligner que le traité sur le cylindre n'a jamais été publié et qu'il nous est parvenu manuscrit. Au contraire, le *De quadratura parabolae* et le *De sphaera et cylindro* ne nous sont parvenus que dans la version imprimée de 1685, tous les manuscrits étant perdus. Il n'existe qu'une version manuscrite d'une proposition du *De sphaera et cylindro* (prop. XXXII *aliter*) qui se trouve maintenant à la Bibliothèque Vittorio Emanuele à Rome.

Nous avons trouvé que dans les *Sereni cylindricorum libelli duo* Maurolico se réfère à un texte sur les sections coniques, qu'il appelle tout simplement *Elementa conicorum* sans nommer l'auteur. Les références à ce texte de conique, très précises et indiquant le livre et le numéro de la proposition employée, nous ont permis de conclure que l'oeuvre citée par Maurolico n'est pas les *Coniques* d'Apollonius[12]. En effet Apollonius n'est jamais cité et, ce qui est bien plus important, les propositions employées par Maurolico sont souvent différentes des théorèmes d'Apollonius. Au contraire dans le *De quadratura parabolae* et le *De sphaera et cylindro* les citations sont explicitement relatives à la reconstruction maurolicienne des *Coniques* complétée en 1547-1548, mais publiée seulement en 1654. Le manque de manuscrits ne permet pas de vérifier si, dans ces oeuvres écrites en 1534, les références aux *Coniques* ont été ajoutées ou modifiées par Maurolico dans la période suivant la rédaction des deux oeuvres. Il nous paraît bien sûr fort improbable que dans deux oeuvres écrites presque en même temps Maurolico cite deux textes différents sur les sections coniques : le premier attribué à Apollonius et l'autre dépourvu de toute référence à son auteur. Il est donc logique de penser que les références aux *Coniques* ont été changées à un moment ultérieur par Maurolico même ou par l'éditeur des deux traités. Cette hypothèse a été confirmée par un folio contenu dans le manuscrit S.P. 115/32 de la Bibliothèque Vittorio Emanuele à Rome. Ce folio, en effet, contient une version extrêmement schématique de la proposition 32 *aliter* du *De sphaera et cylindro*. On a constaté que les références à un texte sur les sections coniques présentes dans ce manuscrit citent les *Ele-*

11. G.B. Memmo, *Apollonii Pergei philosophi mathematicique excellentissimi opera per doctissimum philosophum Iannem Baptistam Memum Patritium Venetum mathematicarumque artium in Urbe Veneta lectorem publicum de Graeco in Latinum et noviter impressa,* Venezia, 1537.

12. Pour une analyse des références aux *Elementa conicorum* présentes dans les *Elementa conicorum* on peut voir : R. Tassora, " I. *Sereni cylindricorum libelli duo* di F. Maurolico e un trattato sconosciuto sulle sezioni coniche ", *op. cit.*

menta conicorum et non pas les *Coniques* d'Apollonius. En ce qui concerne le *De sphaera et cylindro* nous avons donc la preuve que les références présentes dans la version imprimée ont été modifiées à un moment postérieur à la première rédaction. Entre autres, l'une des références dans la version imprimée de cette oeuvre est erronée et coïncide avec la référence présente dans la version manuscrite. Évidemment en modifiant les citations des théorèmes sur les coniques Maurolico n'a corrigé que l'une des deux références contenues dans la proposition.

Nous pouvons constater une situation analogue dans le *De quadratura parabolae* où sont présentes cinq références aux *Coniques* dont l'une est erronée. Pour notre recherche il n'est par particulièrement important d'établir si la citation erronée est due à un simple oubli commis pendant la correction des références à un texte différent : il nous semble, en effet, que les nombreuses citations trouvées dans les *Sereni cylindricorum libelli duo* et les informations offertes par la page manuscrite dont nous venons de parler sont suffisantes pour démontrer qu'en 1534 Maurolico employait un texte sur les sections coniques différent de sa version finale des *Coniques* d'Apollonius : son *Emendatio et restitutio*.

D'autre part ce texte ne peut pas être l'un des textes du Moyen Age sur les sections coniques car il s'agit d'une oeuvre très vaste composée au moins de deux livres dont le premier contient plus de 82 propositions. A présent on ne connaît aucune oeuvre de ce genre dans la tradition du Moyen Age. Voilà pourquoi nous pensons que ce texte de *Elementa conicorum* doit être une oeuvre que Maurolico avait écrite avant 1534, mais qui a été perdue.

Les indications obtenues par une étude attentive des références à notre disposition, compte tenu du développement des études mauroliciennes sur les sections coniques semblent montrer que cette oeuvre de jeunesse sur la section du cône aurait été écrite par Maurolico avant de connaître directement l'oeuvre d'Apollonius. Nous pensons que Maurolico avait reconstruit un texte sur les sections coniques en employant toutes les informations qu'il avait apprises à ce sujet pendant ses premières études. De cette façon pour reconstruire avec cohérence les traités d'Archimède et le traité de Serenus sur le cylindre Maurolico devait disposer d'une théorie sur les coniques, mais en même temps à travers l'étude de ces traités il pouvait comprendre et apprendre certains éléments de cette théorie.

Mes conclusions sur ce que Maurolico a fait avant 1534 sont les suivantes : à travers l'oeuvre de Giorgio Valla et plus tard à travers la lecture du *De quadratura parabolae* et peut-être d'un quelconque traité sur le miroir ardent il a appris certains éléments importants sur la théorie apollonienne. A partir de ces éléments, qui étaient pourtant partiels, Maurolico a reconstruit une théorie qu'il a exposée dans ses *Elementa conicorum*. Dans les oeuvres écrites pendant cette période Maurolico fait référence à ce texte, mais après la publication de la traduction de Memmo des *Coniques* d'Apollonius en 1537 Maurolico a compris que l'oeuvre qu'il venait de reconstruire était plus pauvre que les *Coniques* d'Apollonius et il a décidé d'abandonner son premier écrit pour se consacrer à la reconstruction de l'oeuvre apollonienne. Voilà pourquoi en 1547 il finissait sa version des *Coniques* qui serait devenue le texte de référence en matière de coniques pour toutes les oeuvres mauroliciennes. Mais dans les versions manuscrites des traités de la première période nous pouvons encore trouver les références au texte initial plus simple et pauvre que les *Coniques*, mais pourtant l'une des oeuvres les plus importantes sur les sections coniques de cette période.

Je crois que cette histoire sur les *Elementa conicorum* écrits par le jeune Maurolico peut bien illustrer la méthode de travail de notre mathématicien : face à une lacune dans ses connaissances il cherchait à reconstruire une théorie qui pouvait l'aider à comprendre ce qu'il devait étudier. Dans cette théorie les éléments connus allaient se fondre avec les nouvelles idées et les démonstrations de Maurolico dans une tentative de reconstruction mathématique du savoir des anciens. L'abandon de la part de Maurolico de ses premiers écrits n'a pas été déterminé, j'en suis sûre, par le besoin d'être fidèle à l'oeuvre originale, mais par la considération de ne pas avoir reconstruit tous les théorèmes d'Apollonius et toute sa théorie.

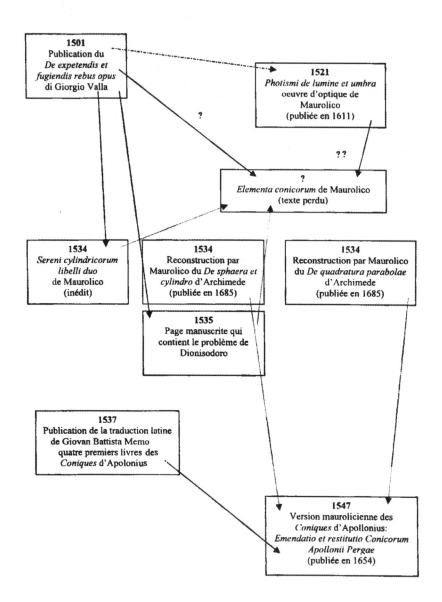

Annexe 1 :

Voici le texte latin de ce projet : *Praesertim, cum sit nobis ad caeteras liberaleis disciplinas transeundum, et praecipue mathematicas : quibus ab ineunte aetate adeo sum oblectatus, ut nullis aliis avidius operam dederim, aut intentius invigilarim. Sed, si quem aditum ad huiusmodi disciplinas habuerim, quantumque, in iis profecerim aedidero : nescio, an futurum sit credibile. Nam universa Elementorum Euclidis volumina, et eiusdem Optica et Catoptrica, nec non Phaenomena et geometrica Data : tamque Iordani, quam Boethi arithmetica et musica Elementa : itemque Theodosii Sphaerica : et quod, omnium maximum est, magnam Ptolomaei constructionem, nullo praeeunte praeceptore, per memetipsum intellexi. Quod si quis non credat periculum faciat, in quocumque voluerit, theoremate vel problemate : cognoscet, me non mentitum. Quid, si maiora his dixero. Vereor ne mendax absque dubio existimer. Dicam tamen : utcunque futurum sit. Quidquid enim Syracusius Archimedes de circuli dimensione : de Sphaera et Cylindro : deque momentis aequalibus disseruit, ego quoque apertissime demonstravi : demonstravi inquam prius, quam ipsius Archimedis opera vidissem. Omnes Menelai de Sphaericis conclusiones ostendi : nec dum Menelai Sphaerica vidi. Verum nullam ex hoc quaero laudem : quid enim feci nisi quod antea factum erat ? Decretum est itaque nobis nonnulla de huiusmodi egregiis disciplinis emittere : ne frustra vigilasse et cum magna valetudinis nostrae iactura laborasse videamur. Sed operae precium est, quando in eum sermonem indicimus, opuscula quaedam a nobis in lucem danda enumerare. Emittenda est in primis ad bonarum artium amatores Epistola : in qua singularum artium mathematicarum, et earundem autorum laudes explicabuntur : et quorumdam audaculorum errores patefient. Deinde libellus de Sphaera mobili : in qua circulorum et arcuum diffinitiones, et circa eosdem omnimoda Theoremata Theodosii Sphaericis inhaerentia disponentur. Tertio dehinc loco, quaedam in Peurbachii Theorias oportunae et necessariae additiones. Post haec astronomicorum problematum libelli quatuor : in quibus omnis Astronomiae calculus geometricis innixus fundamentis explicabitur. Quibus annectetur sinuum sive chordarum Tabella circuli semidiametrum millies mille particularum supponens. His adiiciantur duo libelli, unus de praxis arithmeticae theoria : alter de arithmeticis Datis. His et alii duo, de Photismis unus : alter de Diaphanis. In illo, praeter caetera, patescet cur solaris radius per qualecumque foramen transmissus in circularem redigatur formam: in hoc ratio rotunditatis et colorum Iridis aperietur : Quorum utrunque fuit Ioanni vulgatae Perspectivae authori incognitum. Ad haec quosdam locos annotabimus circa linearum Symmetriam : circa Solidorum structuram : circa maximas planetarum aequationes : ubi ostendemus, quod circulo, cuius dimetiens rationalis supponatur, Octogonum atque Dodecagonum aequilaterum circumscribente : ipsius octogoni latus erit irrationalis linea, que Minor : Dodecagoni vero latus, ea, quae Apotome a praestantissimis mathematicis appellatur. Determinabimus etiam locum lunaris deferentis : in quo maxima centri contingat Aequatio : ubi turpiter erravit is, qui planetarum Theorias exposuit. Postremo dabitur totius mathematicae disciplinae Compendium quoddam ex Euclide, Theodosio Archimede, Menelao, Iordano, Boetho (sic), Ptolomaeo, caeterisque acutissimis mathematicis excerptum. Quibus ex rebus apparebit, quantum et laboraverim et profecerim. F. Maurolico, Grammaticorum rudimentorum libelli sex, 7–8.*

Annexe II :

Voilà un extrait de la lettre de 1536 au Cardinal P. Bembo : *Ego quidem in bonis artibus et mathematicis praecipue disciplinis diu versatus sum : quarum me quandoque tantus amor occupavit, ut caeteras philosophiae partes pene contempserim. Sed illud non aegre ferre nequeo, quod tales hodie disciplinae non ita, ut decet, excoluntur. Floret ubique Galenus : resonant Academiae Iustiniano : rumpuntur marmora dialecticis iurgiis. Cur silet Euclides praestantissimus ? Cur silet Archimedes ac Theodosius ? Cur Menelai, Apollonii, Sereni Praeclara nusquam audiuntur nomina ? An ideo quoniam Galenus et Iustinianus soli, ut aiunt vulgo, quaestum et vitae commodum afferant, in caeteris nihil frugis subsidat ? Vulgaris est haec sententia et imprudenter prolata. Nam si recte considere s, caeterae disciplinae sine mathematicis esse aut certe bene esse non possunt. [...] Ex Euclide vix sex elementorum libri leguntur : ad caeteros nemo progreditur, ceu ad ignotas oras. Nam Theodosii et Menelai Sphaerica, Apollonii Conica, Archimedis opera de circuli dimensione, De Sphaera et Cylindro, de Isoperimetris, de momentis aequalibus, de Quadratura parabolae, de Speculis ignificis nusquam apparent, non secus ac si admisso inexpiabili perpetuum meruerint exilium. Et horum si quid circumfertur, tot tantisque scatet mendis, ut vix etiam ab authore ipso emendari possit. Astronomica quoque studia adeo exoleverunt, ut Ptolemaeo, caeterisque optimis authoribus neglectis, nil nisi Sphaeram Ioannis de Sacro Bosco legamus, proponamus, celebremus, quasi opus egregium et notatu dignum. Sed quid mirum ? Nemo potest hominis illius errores deprehendere, nisi Ptolemaeum praelegerit. Quod si Ioannes a Regio Monte aut eius praeceptor Georgius Peurbachius, uterque mathematicus consumatissimus, sicut planetarum theorias, sic Sphaerae rudimenta nobis edidisse iam pridem cum Gerardo Cremonensi una Ioannes hic de Sacro Bosco rudis astronomus exsibilatus et explosus fuisset.*

[...] Libet mihi, Bembe vir doctissime, de tanta Doctrinae huius calamitate tecum conqueri : qui omnis generis scientiarum et amator et defensor es eximius. Ego, quantum mihi licuit, assiduis studiis, mentisque agitationibus, nisus sum collatis priscis exemplaribus et dictorum authorum et aliorum opera complura emaculare, et in suum restituere nitorem ; quae tibi gratissima fore spero, si quando curis necesariis, laboribusque vacuus his vacare poteris. Scripsi quoque per me nonnulla videlicet de figuris locum implentibus ubi Averrois error patebit, qui putavit sicut cubos inter regularia quinque solida, ita et pyramides per se locum implere. Item Sphaericorum libellos. v. Arithmeticam, Arithmetica data : in quibus multa a Boethio, Iordanoque praetermissa, demonstrantur. Photismos, in quibus solares radii per qualevis foramen transmissi rotunditas demonstratur. Diaphana, in quibus multa de iride, quae nec ubi leguntur. De motuum symmetria. Arithmeticas, geometricasque quaestiones. De Sphaera mobili. Speculationes multas, ubi inter caetera, demonstramus latus octogoni lineam esse minorem, dodecagoni vero latus apotomen, circulo in quam rationalem habente diametrum inscriptorum, et alia complura. Dialogos quoque tres, primum de forma, numero et ordine tam elementorum, quam coelorum. Alterum de circulis Sphaerae et planetarum motibus. Tertio de anno et reliquis temporum spaciis. Quos dialogos, quoniam mundi formam et ordinem continent, placuit appellare Cosmographiam : quam ego tibi, si lubet, dedicare decrevi. Est mihi sphaericum instrumentum circulis aeneis compactum: quod tu olim Venetiis ad Costantinum tuum Lascarem misisti. Sunt in eius zodiaco signa latine graeceque conscripta. Illud in Cosmographia nostra commendatur et explicatur.

Maurolico et Archimède : sources et datation du premier livre du De momentis aequalibus

Enrico Giusti

Le De momentis aequalibus

Les études sur Maurolico ont depuis longtemps mis en évidence de nombreux cas dans lesquels celui-ci n'hésite pas à intervenir sur les textes des géomètres grecs qu'il étudie : il n'hésite pas à changer l'ordre et le développement des démonstrations, à insérer de nouvelles propositions ou à en reconstruire de fond en comble sur des bases nouvelles. Cette manière de procéder est flagrante si l'on examine certains textes d'Archimède[1], et plus particulièrement les quatre livres dont est composé le remaniement de Maurolico de l'*Équilibre des figures planes* (que nous abrégeons dans la suite *EFP*) : le *De momentis aequalibus* (*DMÆ*).

Aucun manuscrit du texte ne nous est parvenu, et l'analyse des sources, le moment de son élaboration et la stratification textuelle interne de l'oeuvre, doivent alors se fonder presque exclusivement sur une analyse de la structure mathématique et sur quelques rares renseignements externes. L'unique texte dont nous disposons est l'édition de Palerme de 1685[2]. D'après les dates indiquées dans celle-ci, les quatre livres du *DMÆ* furent composés entre le 16 décembre 1547 et le 23 janvier 1548. Le premier livre traite du centre de gravité, de la loi du levier, du moment et de la statique ; le second du centre de gravité du triangle et du trapèze ; le troisième du centre de gravité d'un segment de parabole ; le dernier des solides. A peine plus d'un mois pour l'élaboration d'une oeuvre aussi complexe incluant un nombre considérable de

1. *Cf.* M. Clagett, *Archimedes in the Middle Ages*, vol. III, III.

2. *Admirandi Archimedis siracusani Monumenta omnia°. ex traditione F. Maurolyci*, Palermo, Cillenio Esperio, 1685. On peut maintenant consulter cette édition sur le site internet : <http://www.dm.unipi.it/pages/maurolic/index.htm>. Pour l'histoire de l'édition, voir R. Moscheo, " L'Archimede del Maurolico ", in C. Dollo (ed.), *Archimede. Mito, tradizione e scienza*, Firenze, Olschki, 1992, 111–164.

nouvelles démonstrations, est un intervalle de temps peu crédible : ces dates doivent être considérées comme celles de l'écriture sur le papier de la version définitive du texte[3].

Nous savons par ailleurs qu'au début des années 1530, Maurolico avait écrit une version aujourd'hui perdue du *DMÆ*. Dans un manuscrit autographe daté de 1534, Maurolico fait une référence explicite à une proposition d'un *libellus momentorum aequalium*, précisément à la proposition aujourd'hui numérotée 5 du premier livre de la version de 1547-1548[4]. Il déclarait aussi dès 1528 : *Quidquid enim Syracusius Archimedes de circuli dimensione, de sphaera et cylindro, deque momentis aequalibus disseruit, ego quoque apertissime demonstravi, demonstravi inquam prius quam ipsius Archimedis opera vidissem. Omnes Menelai de Sphaericis conclusiones ostendi : nec dum Menelai Sphaerica vidi*[5].

L'affirmation est étonnante : est-il possible de reconstruire une oeuvre " sans l'avoir jamais vue " ? L'affirmation est aussi mystérieuse : si on juxtapose le passage suivant sur Ménélaus sans *nec dum*, on pourrait conclure qu'au moment où il écrit ces lignes, Maurolico disposait alors d'un texte d'Archimède. Mais lequel ? En 1528, il n'existait aucune édition imprimée des oeuvres d'Archimède contenant *EFP* ou *La sphère et le cylindre*[6]. Et qu'avait donc reconstruit Maurolico en 1528 d'Archimède ?

Ces questions complexes d'identification des sources et des travaux archimédiens de Maurolico n'ont pas été encore complètement résolues. Nous nous limitons à seulement signaler les deux dernières, mais nous consacrerons le reste de cet article à la première : nous essaierons de déterminer si, et si oui dans quelle mesure, Maurolico a pu obtenir les résultats archimédiens dont il parle sans avoir vu l'oeuvre d'Archimède. De quelles sources s'inspira-t-il alors ? La teneur du texte de 1528 nous est inconnue, mais devait contenir au moins les fondements qui représentent une partie importante du premier livre du *DMÆ*. L'analyse de ce dernier texte sur lequel nous avons concentré nos efforts nous fournira déjà quelques résultats.

---

3. De nombreux indices laissent d'ailleurs à croire que Maurolico a encore quelquefois lourdement retouché son texte après 1548 : cf. P.D. Napolitani et J.-P. Sutto, *F. Maurolico et le conoïde parabolique*, à paraître.

4. M. Clagett, *Archimedes in the Middle Ages*, vol. III, *op. cit.*, 885, ligne 44.

5. *Grammaticorum libelli sex*, Messina, P. Spira, 1528, ff. 7r–7v.

6. L'unique texte imprimé disponible était le *Tetragonismus* publié à Venise en 1503. Il contient les traductions de Moerbeke de la *Quadrature de la parabole* et de la *Mesure du cercle*. Clagett a démontré que les éditions de Maurolico de ces deux textes étaient de fait basées sur l'édition de Gaurico.

LE *DE MOMENTIS AEQUALIBUS* ET GIORGO VALLA

De quelles sources Maurolico pouvait-il disposer à cette époque ?

1. La version originale de *EFP*.

2. " L'encyclopédie " *De expetendis et fugiendis rebus* de Giorgio Valla, publiée à Venise en 1501. On a pu montrer que Maurolico en fit une utilisation intense[7], et elle contient quelques passages du commentaire d'Eutocius à *EFP*.

3. Des sources non classiques ni même archimédiennes, en particulier la " science des poids " et d'éventuelles autres traditions médiévales.

Examinons de plus près quels éléments Maurolico pouvait puiser dans Valla. Ce dernier possédait le codex A d'Archimède dont il traduisit quelques passages. Un premier élément permet de rapprocher Maurolico et Valla. Le titre dans le *De expetendis et fugiendis rebus* par lequel est traduit *ΕΠΙΠΕΔΩΝ ΙΣΟΡΡΟΠΩΝ* est *Liber qui aequalium momentorum inscribitur*. La traduction de Moerbeke avait *Liber Archimedis de centris gravium vel de planis aequerepentibus*, et celle de Jacques de Crémone *Archimedis planorum aequeponderantium inventa vel centra gravitatis planorum*. Le texte traduit par Valla consiste seulement en quelques passages tirés du commentaire d'Eutocius : *Archimedes autem in libro qui aequalium momentorum inscribitur centrum momenti planae figurae putat a quo elevatus manet finienti parallelus, duum aut plurium planorum momenti centrum aut gravitatis, unde lanx elevata parallelus est finienti.*

Ut si sit triangulum ABC, in medioque ipsius punctum D, a quo elevatus manet parallelus finienti ; non dubium putat quin aequilibrae sint A,B,C, partes sibi, neque altera magis altera rhepet in finientem. Itidem lance existente AB et sublatis ex ipsa A, B magnitudinibus, si lanx sustolletur a c, aequilibras habebit A, B partes parallelus manent finienti, eritque centrum elevationis ipsarum A, B magnitudinum c.

" Aequalibus et similibus ", inquit Geminus cum Archimede, *" planis figuris accomodantibus invicem etiam centra gravitatum invicem sese accomodant. "* *Nam omnes partes omnibus consentiunt. " Inaequalium autem at similium centra gravitatum similiter erunt affecta ".*

Nous allons examiner si les différentes sources que nous avons énumérées, en particulier ces fragments traduits par Valla, ont trouvé place dans l'élaboration du premier livre du *DMÆ*, et si oui, de quelle manière.

---

7. Voir la contribution de R. Tassora dans ce volume. Tous les passages relatifs à Archimède dans Valla sont publiés dans M. Clagett, *Archimedes in the Middle Ages*, vol. III, 474-475.

Les definitions et postulats

La version de Maurolico commence par une série de 13 définitions et 8 postulats. Nous les avons classés *a priori* selon leur source probable. La division met en évidence que pour toute une série de définitions et postulats, Maurolico a pu *effectivement* utiliser une ou des sources différentes du texte de *EFP*. Inversement la quatrième section (Archimède) contient les postulats que nous n'avons pu trouver que dans *EFP*. d désigne une définition et p un postulat ; la numérotation suit celle du texte.

La tradition non-archimédienne

d1. *Universale centrum est punctum, in quod unumquodque gravium naturaliter, et recto motu tendit.*

d2. *Horizon est planum, cui perpendicularis est recta, quae a puncto quopiam in centrum universale ducitur.*

d3. *Magnitudo rei est eius dimensio iuxta unam, vel duas, vel tres longitudines.*

d4. *Pondus rei est gravitas nitentis ad centrum.*

d5. *Unde magnitudinum aequalium pondera interdum sunt inaequalia ; et contrario ponderum aequalium magnitudines quandoque sunt inaequales.*

d6. *Centrum magnitudinis est punctum aequaliter remotum ab extremis ; ut centrum circuli, aut sphaerae, aut figurae solidi regularis.*

d7. *Centrum gravitatis est punctum, in quod gravi undecunque suspenso, a signo suspensionis acta linea horizonti perpendicularis est.*

d13. *Signum autem appensionis aequilibrii punctum vocatur.*

p1. *Grave quodpiam a punto quolibet suspendere, ut libere pendeat.*

p2. *Centrum gravitatis ad centrum universale, quantum possibile est, proxime accedere.*

Eutocius–Valla

d11. *Gravia vero aeque pendere seu aeque ponderare dicuntur, cum ab aliquo puncto appensa ita pendent, ut recta quae gravitatum centra, vel appensionum puncta coniungit, horizonti aequedistet.*

p6. *Figurarum sibi invicem congruentium congruere et gravitatis centra.*

p7. *Figurarum similium centra similiter posita esse : hoc est quemadmodum figurae similes sunt, ita gravitatum centra cum terminis respondentium connexa laterum similia conficere triangula.*

Moment

d8. *Momentum est vis ponderis a spacio quopiam contra pendentis.*

d9. *Unde ponderum aequalium momenta possunt esse inaequalia ; et e contrario contiget momentorum aequalium pondera esse inaequalia.*

d10. *Aequalia enim momenta sunt gravium aeque ponderantium, sive aequependentium.*

d12. *Maius autem momentum est ponderis, quod deorsum inclinatur.*

Archimède

p3. *Si gravium aeque pendentium alteri quid adiectum sit ; idem deorsum inclinari, reliquum sursum ferri [post. 2 EP].*

p4. *Contra si gravium aeque ponderantium uni quid auferatur ; idem sursum, reliquum deorsum ferri. [post. 3 EP].*

p5. *Si duo gravia aeque ponderent et duo quaelibet eis aequalia singula singulis ad eadem spatia aeque ponderabunt [post. 6 EP].*

p8. *Centrum gravitatis esse intra rei gravis ambitum [post. 7 EP].*

Notons que certains postulats et définitions pourraient éventuellement appartenir à plusieurs groupes. C'est le cas par exemple, de d6 (centre de gravité). De plus, la section " Moment ", semble être une création ex novo du mathématicien de Messine. Non pas que le terme n'ait pas d'histoire[8], mais Maurolico l'a pris probablement chez Giorgio Valla. C'est toutefois la première fois que nous en trouvons une définition précise. Qui plus est, à partir de la définition d10 et d12, il devient possible de comparer des moments et donc d'utiliser ce concept comme une grandeur obéissant à la théorie des proportions.

LE RÔLE DES AXIOMES DANS LES DÉMONSTRATIONS

La première partie du texte peut être mise entièrement sous le signe de la tradition non archimédienne. Elle est dédiée à deux problèmes : la détermination mécanique du centre de gravité, par la suspension du grave en deux points différents (prop. 1–5) ; et la démonstration que le centre de gravité et le point d'équilibre coïncident (voir prop. 11 et d13). Ces propositions se basent uniquement sur les sections " médiévales " du système d'axiomes et de définitions, à l'exception de la proposition 11 qui nécessite de savoir que le centre de gravité d'un système de deux corps se trouve sur la droite qui joint les centres de gravité (prop. 6).

Et c'est justement avec la proposition 6 que commence la partie " archimédienne " du traité. p6 entre ici en jeu. Ce postulat archimédien, que Maurolico a pu tirer de Valla, se révèle être le principal acteur de ce livre. Examinons la démonstration. Soient les centres de gravité des deux graves A et B, a. et b., et le centre de gravité du système composé de ces deux graves, g. Par

8. Voir P. Galluzzi, *Momento*, Roma, Ateneo e Bizzarri, 1979.

un raisonnement par l'absurde, supposons que g. se trouve en dehors de la droite a.b. Considérons un second système de graves, D et E, de centres d. et e., congruents à A et B. Si nous superposons D à A et E à B, le centre de gravité du système D+E se superpose à g. par p6. Mais, si avant de les superposer, nous tournons les poids D et E de 180° autour de l'axe d.e., le centre de gravité de D+E correspond alors avec un autre point, q., différent de g., ce qui contredit p6.

p6, disions-nous, a un rôle central dans la démonstration des principaux résultats du premier livre du *DMÆ* : équilibre de poids égaux, centre de gravité du rectangle, loi du levier, etc. Le lecteur aura en effet noté — peut-être avec quelque étonnement — qu'à la différence d'Archimède, Maurolico ne postule pas l'équilibre de la balance ayant des bras égaux auxquels sont suspendus des poids égaux ; de même dans les axiomes, on cherchera en vain une situation où est supposé l'équilibre.

Examinons comment alors Maurolico obtient le résultat : *Centra gravium aequalium, aequaliter distant a communi centrum* (prop. 16). Soient A et B deux poids égaux et soit g. leur centre de gravité. Considérons deux autres poids D et E, égaux à A et B, placés à la même distance et soit q. leur centre de gravité. Si nous superposons D à A et E à B, par p6, les centres coïet on aura a.g.=d.q. Si l'on superpose au contraire D à B et E à A, nous obtenons b.g.=d.q., et donc a.g.=b.g., cqfd. Dans cette démonstration aussi, le centre est déterminé par la symétrie de la configuration géométrique.

Le même mécanisme joue dans la proposition 25. Maurolico y détermine le centre de gravité d'un grave " uniforme " : un rectangle ou un parallélépipède. On en déduit facilement la loi du levier, moyennant la réduction des deux graves à un unique rectangle ou parallélépipède. La méthode rappelle fortement celle qu'utilisait Archimède dans la première proposition du second livre de *EFP*.

Ces exemples suffiront pour comprendre le rôle crucial que p6 joue dans la théorie du centre de gravité de Maurolico. Il permet d'obtenir des informations sur le centre de gravité d'un système en exploitant la symétrie des configurations. On notera surtout que tous les résultats exposés jusqu'ici ne dépendent que de postulats et de définitions que Maurolico pouvait trouver en dehors de *EFP*.

LE MOMENT ET LA STATIQUE

Maurolico avait déjà dans les définitions, lié moment et équilibre. Mais l'équilibre est une notion qualitative : un équilibre se maintient, ou se rompt ; on ne peut rien dire de la quantité dont un équilibre dépasse l'autre. Le *moment* de Maurolico au contraire, est une grandeur géométrique : les moments peuvent être comparés, ils peuvent aussi s'ajouter et figurer dans des rapports. Le moment traduit quantitativement l'équilibre.

Pour qu'il soit possible de considérer le moment comme une grandeur, obéissant aux lois de la théorie des proportions, il est nécessaire que l'on puisse construire des multiples quelconques d'un moment donné. Maurolico en est conscient, mais ne semble pas juger nécessaire une axiomatique spécifique. Il essaie de démontrer la relation entre poids et moment pour une distance fixe (prop. 36) : si le poids suspendu à une certaine distance est doublé, le moment est aussi doublé, etc. La démonstration, faute d'axiomes explicites, devient un cercle vicieux ; mais, une fois cette relation établie, le traité continue sur un mode parfaitement rigoureux. Maurolico démontre — suivant un parcours qui deviendra classique après Galilée — que le rapport des moments se compose du rapport des graves et de celui des distances auxquelles les graves sont suspendus (prop. 39)[9] :

$$\frac{M_1}{M_2} = \frac{P_1}{P_2} \otimes \frac{d_1}{d_2}$$

La théorie des moments de Maurolico est ensuite immédiatement appliquée à l'étude de la statique " réelle ". L'équilibre y est déterminée par les poids placés sur les plateaux et par la position du contrepoids, et par le poids et la position des parties constitutives de la balance. Il nous est ici impossible de rendre compte de cette application, courante dans la science médiévale des poids (par exemple dans le *De canonio*) ; nous nous bornerons à remarquer que Maurolico, à la différence de ce que l'on trouve dans la tradition médiévale, construit tout son développement autour de l'utilisation du moment et de la relation obtenue plus haut entre poids, moments et distances.

Le premier livre du *DMÆ.* se conclut sur cette application, et on notera de nouveau que tous les moyens conceptuels auxquels Maurolico aura eu recours ne dépendent pas nécessairement d'une lecture de *EFP*.

CONCLUSION

Dans tous les exemples qui précèdent, nous avons vu comment Maurolico pouvait avoir utilisé de façon différente les sources qui étaient alors disponibles. Seul le thème de la statique rapproche le mathématicien de Messine de la science des poids. Des autres traditions médiévales — un travail approfondi de recherche de ces sources potentielles est absolument nécessaire —, dérivent certains concepts " cosmologiques ", la définition du centre de gravité et son identification avec le point d'équilibre du levier. Il s'agit sans aucun doute d'un point significatif, puisqu'il permet à Maurolico une organisation conceptuelle qui dans le texte original de *EFP*, reste dans l'ombre. Mais le rôle de la tradition archimédienne s'arrête ici. Dans le reste du traité, seules restent nécessaires les

---

9. Nous renvoyons sur ce point à P.D. Napolitani, " La géométrisation de la réalité physique ", in P. Radelet et E. Benvenuto (eds), *Entre mécanique et architecture*, Basel, Birckhäuser, 1995, dans lequel est exposée en détail la théorie de Maurolico du moment ; voir aussi E. Giusti, " Ricerche galileiane : il *De motu aequabili* ", *Boll. St. Sc. Mat.*, 6 (1986), 89-108.

deux idées suivantes : la possibilité de supposer connu le centre de gravité, et celle de l'identifier avec le point d'équilibre.

Le rôle du matériel que Maurolico prend de Valla, la notion de moment et le " principe de symétrie " véhiculés par p6, est bien différent. Le lecteur objectera que ces notions sont à peine perceptibles dans Valla. Cela fait d'autant plus ressortir, répondrons-nous, l'importance de la réorganisation et l'aménagement de ces concepts par Maurolico. On peut mesurer la libéralité du mathématicien messinois devant des sources de toute façon difficilement utilisables. Maurolico tire la notion de moment de la terminologie du *De expetendis*, et la transforme en une grandeur géométrique arbitrant l'équilibre du levier. Sa théorie des moments est probablement le premier exemple à l'époque moderne de géométrisation d'une grandeur physique. Le principe de symétrie quant à lui, est une méthode démonstrative qui devient la clef des démonstrations cruciales de *DMÆ*.

Quant aux thèmes archimédiens que Maurolico n'aurait pu tirer que de *EFP*, leur présence dans le premier livre est à peine perceptible. Cette situation particulière ne se répétera pas dans les livres suivants. Si on y traite encore du centre de gravité des figures, les postulats archimédiens, et en particulier p8 reprennent leur rôle central, même s'il est tempéré par l'utilisation constante du principe de symétrie. Une analyse des deuxième et quatrième livres (le troisième est, de l'aveu même de Maurolico, inspiré par une lecture du second de *EFP*) pour discerner si, et si oui de quelle nature, il a pu exister une couche de ces textes indépendante de la lecture directe d'Archimède, sort des limites de ce travail.

Est-il possible que Maurolico ait construit le premier livre du *De momentis* sans avoir jamais vu un texte d'Archimède ? La réponse semble donc devoir être affirmative. Nous pouvons conclure que nous sommes en présence d'un texte essentiellement non archimédien, composé au moins dans ses grandes lignes avant 1528, sans que le mathématicien ait eu à disposition l'*Équilibre des figures planes*.

FRANCESCO MAUROLICO'S EDITION OF THE *CONICS*

Ken SAITO

EMENDATIO OF THE *CONICS* BY MAUROLICO

Francesco Maurolico made an edition of Apollonius' *Conics*, titled *Emendatio et Restitutio Conicorum Apollonii Pergaei*, which remained unpublished until 1654[1].

Maurolico knew very little about the theory of conic sections before 1537, when Memo's latin translation of Apollonius' *Conics*, which is the base of Maurolico's emendation, was published for the first time[2].

A manuscript of 1534, Maurolico's autograph[3], which is an adaptation of Serenus' *Section of cylinder and cone* shows Maurolico's poor knowledge of conic sections at this time[4] . We can conclude that what little Maurolico knew about the conics came from Giorgio Valla's *De expetendis et fugiendibus rebus opus* (1501) and some other sources he came across, which contained no more than rudimentary propositions and properties about conic sections.

Memo's translation, however imperfect and full of errors it may have been, showed a theory of conic sections much more developed than Maurolico could

---

1. *Francisci Maurolyci Messanensis, Emendatio, et Restitutio Conicorum Apollonii Pergaei …*, Messanae, 1654. This edition also contains Maurolico's original reconstruction of fifth and sixth books of the *Conics*. For the contents of these posterior books see A. Brigaglia's paper in this volume and his " La ricostruzione dei libri V e VI delle *Coniche* da parte di Francesco Maurolico ", *Bollettino di storia delle scienze matematiche*, 17 (1997), 267-307. The autograph manuscript of the first four books is extant : Real Biblioteca de San Lorenzo de el Escorial, ms J.III.31. For the restauration of the posterior books, R. Moscheo has recently found a manuscript in the archive of the Pontifica Università Gregoriana (Rome) by the hand of C. Grienberger, jesuit, student and successor of C. Clavius. However, the autograph has not yet been found.

2. *Apollonii Pergei philosophi, mathematicique excellentissimi Opera per doctissimum philosophum Ioannem Baptistam Memum … de Græco in Latinum et noviter impressa*, Venetiis, 1537.

3. *Sereni cylindricorum libelli duo*, Bibliothèque Nationale de Paris, Fonds Latin 7465.

4. For details, see R. Tassora's article in this same volume, and her " I. *Sereni Cylindricorum libelli duo* di F. Maurolico e un trattato sconosciuto sulle sezioni coniche ", *Bollettino di storia delle scienze matematiche*, 15 (1995), 135-264.

imagine from the sources at his disposal. It was almost logical that the imperfection of Memo's translation led Maurolico to embark upon its improvement. This in fact is what Maurolico did in the first half of 1547, in a fairly short period. For example, according to the colophon of the manuscript, he finished the third Book in the night of June 2nd, and the fourth on 24th of the same month.

Maurolico's edition can be summarized as a mathematico-logical emendation of Memo's translation. Philological fidelity was not among Maurolico's goals, as often attested in his works. His text contains a lot of additions and modifications, in order to " improve " Memo's version from mathematical point of view.

Maurolico's diligence and competence in this kind of work is really impressive. No errors, no lacunae of whatever kind in Memo's edition, from simple typographical error until omission of a couple of lines that made the text unintelligible, could escape Maurolico's eyes ; all of them are corrected in some way.

Many references to the propositions of Euclid's *Elements*, and to those of the *Conics* itself, are added. Not only did he made additions to the text, but was also ready to delete arguments which seemed to him obvious and trivial, and to replace awkward demonstrations by his own. To give you an idea of his way of work, let me cite a couple of examples. The first one is very simple : in the proposition 5 of the first book, Memo confound the relative pronoun *hou* (of which) with very common adverb of negation *ou* (not), and reduces the text unintelligible. This is surely one of the errors due to Memo himself (some other errors can be ascribed to the editor of the posthumous edition).

Heiberg 1:18,21 : *kuklos ara estin, hou diametros hê* DE.

Memo 4v,15 : *circulus igitur est non diameter linea* DE.

Maurolico 9,18 : *quare per praecedentem Planum* EZT *circulum, qui est* ETD *in cono.*

Maurolico adds *per praecedentem*, that is, " by the previous proposition I-4 ", and invents a phrase, which is not mathematically wrong, but he did not succeed to restore the sense of the original text of Apollonius. This suggests that Maurolico did not have other sources of the *Conics* at least while he was working on Memo's translation[5].

Let us see another example which enables us to conclude that Memo's book was the only source for Maurolico's version of the *Conics*, and that the latter was little interested in restoring Apollonius words. What we are going to examine is a lacuna in Memo's edition, and as there is no problem about translation, I cite it by Heiberg's latin translation. In the protasis of proposition I-

13, Apollonius describes a ratio of the square on AK to the rectangle contained by BK e KΓ :

Heiberg 49,12 : *[rationem] ..., quam habet quadratum rectae [= AK] a uertice [= A] coni diametro sectionis parallelae ductae usque ad basim trianguli [= triangle ABG] ad rectangulum comprehensum rectis ab ea ad latera trianguli abscisis [= rectangle BK, KΓ], latitudinem habens ...*

Memo 9v,4 : *[rationem] ... quam quadratum quod fit a ducta sumitate Coni preter diametrum sectionis usque ad basim trianguli latitudinem habens ...*

Maurolico 14,1 from the bottom : *[rationem] ... quam quadratum, quod fit a ducta a vertice coni penes diametrum sectioni usque ad occursum basis trianguli, ad contentum sub tota linea (quae constat ex basi, et adiuncta occurrente) et sub ipsa occurrente ; et latitudinem habens.*

The word *trianguli* which appears twice refers to the triangle ABΓ. Memo jumps from the first *trianguli* to the second, so that his text is completely unintelligible. Maurolico's version restores the mathematical sense of the text, but the use of parenthesis and the word *occurrente* which never appears in other places (as far as I know), show clearly that Mautolico did not use other sources than Memo, and that after all he was not so much concerned in finding an expression that would be nearer to that of Apollonius. Besides, Maurolico's dependence to Memo is also obvious from similar expressions such as *quadratum quod fit a ducta*.

One could cite dozens of other examples of this kind of Maurolico's emendations (in other words, Memo's edition is so full of errors)[6], while I have not encountered even one phrase which Maurolico took from ancient materials independent from Memo (Apollonius's text, Eutocius' commentary or Pappus' lemmata), to correct the errors found in Memo's translation. Therefore it is quite certain that Maurolico depended only on Memo and on his own mathematical talent for his *emendatio* of the *Conics*.

5. Some ten lines afterward, at the end of the same proposition, [Heiberg, 1:20,7], Memo repeats the same error : *non diameter*, where Maurolico finds the correct translation : *cuius diameter*. This does not necessarily suggest that Maurolico consulted some other source, since it is easy to find the correct translation by conjecture. This example rather suggests that Maurolico worked in a hurry, without losing time returning back to the precedent parts of the text. It should be added that often Memo also translates correctly *hou* by *cuius* (e.g., I-12, Heiberg, 1:44,24 = Memo 7v,2). We may assume that Memo did not have the time to reexamine his own translation which was published only posthumously.

6. Some other examples are analyzed in my " Quelques observations sur l'édition des *Coniques* d'Apollonius de F. Maurolico ", *Bollettino di storia delle scienze matematiche*, 14 (1994), 239-258.

BOOK 4 : NUMBER OF POINTS OF INTERSECTION AND CONTACT
OF TWO CONIC SECTIONS

Around the end of Book 4 of the *Conics*, Apollonius confronts the problem
of the maximum number of the points in which two conic sections can intersect or touch each other. For example, the last three propositions establish :

IV—55 Opposite sections [hyperbolas with two branches] do not intersect in
more than four points.

IV—56 If opposite sections touches each other in one point, they do not intersect in more than two points.

IV—57 If opposite sections touches each other in two points, they do not intersect in other points.

For us moderns, the problem is simple, since we can reduce it to that of the
number of solutions of a system of equations representing conic sections.
However, it is not so simple for those who lived before Descartes. In fact,
Apollonius' text itself, as it comes down to us, does contain some errors,
though his arguments furnishes a sufficient ground to demonstrate any possible
case, and therefore we can conclude that Apollonius essentially resolved the
problem.

Maurolico's version of these propositions shows that he understood Apollonius' argumentation very well, and that he had examined all the propositions.
For example, where Apollonius is in error (IV-43 = IV-37 in Maurolico), Maurolico gives his own proof. Also in other propositions, he sometimes replaces
Apollonius' argument by his own.

I concentrate here in a small confusion in Maurolico's manuscript and in the
printed edition of 1654, which allows us to see his way of work. We will also
see that the process of printing in 1654 was not so accurate[7].

After the proposition IV-42 (= Memo IV-47 = Heiberg IV-49), Maurolico, in
his manuscript, went on to the next proposition which, of course, would have
to be the IV-43. However, this proposition became IV-44, because another
proposition was inserted. The proposition at issue is as follows : Maurolico
(manuscript) : IV-44 = Maurolico (print) : IV-43 = Memo : IV-48, Heiberg : IV-
50.

*Si hyperbole alteram contrapositarum ad unum punctum tangat,
contraposita ipsius alteri contrapositarum non coincidet ad plura puncta,
quam duo ; coincidet inquam secando ; vel ad unum tangendo.*

7. I owe this part of my communication to B. Gastardelli and V. Lombardi, undergraduate students at Università di Pisa, who have presented their study on Maurolico's fourth book in a
seminar.

Here Maurolico comes away from Apollonius' argument who uses the properties of hyperbolas and opposite sections proved in previous books of the *Conics*, trying instead to prove the proposition from the results about the number of points of intersection established in Book 4. This choice made it necessary to use a proposition which, in Apollonius's text, appears later, and Maurolico inserted just after his IV-42 (Memo, IV-51 = Heiberg, IV-53 = Maurolico (manuscript) IV-43 = Maurolico (print) IV-44). However, Maurolico seems to have realized its necessity only when he began to write his proof of IV-44 (in manuscript). In fact, this proposition 44 follows directly the 42 in the manuscript, in the verso page of the same folio[8].

We can reconstruct what has happened as follows : Maurolico began to write his proposition 44 (at this point 43) after 42. At some moment, he found that a proposition which had not yet been proved (Memo IV-51 = Heiberg IV-53) was necessary, and he decided to bring this before the proposition he was writing. But as he did not want to rewrite what had already been written, probably also because he was writing the verso side of a sheet, he added this inserted proposition (Maurolico manuscript IV-43) afterwards, in the next sheet.

This conjecture can also be confirmed by the text of the proposition 44 (manuscript). The proof consists of five cases ; the first uses IV-38 (Maurolico's numbering), the second 40, the third 41, the fourth 42. This proposition 42 is referred to as *per praecedentem 42^{am}*. By the expression *praecedentem* Maurolico usually indicates the proposition immediately before. So at this point Maurolico thought to be writing the proposition 43, not 44. Then, in the fifth and the last case, where he had to recourse to the inserted proposition, he wrote : *per immediate praemissam*, an unusual expression in Maurolico, which indicates the proposition 43. This strongly suggests the insertion of the proposition 43. Unfortunately, the printed edition of 1654 follows automatically the order of the manuscript, without noticing the exchanged numbers of 43 and 44.

This example allows us to see again that the manuscript is not so much an elaborated work than a first draft.

BETWEEN AUTOGRAPH (1547) AND PRINT (1654)

The complicated situation in the manuscript we have seen above induces us to ask about the relation between manuscript and the printed edition. This question still requires much profound researches and it is quite audacious to try to give it an answer. Here I try to indicate some possibility following the analysis of Pier Daniele Napolitani.

8. As the manuscript is bound on the left side of the paper, the numbers of some propositions cannot be read, at least in microfilm. One can nontheless safely guess the numbers of these propositions from those of other propositions.

The most simple assumption that the printed edition of 1654 is based on the autograph we possess, turns out to be improbable, because of the discrepancy between them. Besides, we know (thanks to the research of Napolitani and his colleagues) that there existed a certain number of copies of Maurolico's *Conics*, not only for the first four books but also for the Books V and VI, and they were circulated especially among the Jesuits. That the printed edition is based on some source considerably different from the autograph is also supported by Grienberger's copy of Books V and VI, which has recently been found by Moscheo ; according to Napolitani, the source of this copy seems to be nearer to Maurolico's original than that of printed edition.

The problem is no less complicated for the first four books for which we possess the autograph, since one cannot exclude the possibility that Maurolico had it copied and continued to work on this copy (remember that the autograph in 1547 has characteristics of a first draft, and there certainly existed several copies other than autograph). Then, it is possible that in some cases some apograph or printed edition should reflect Maurolico's mature thoughts.

We (the group that has organized this symposium on Maurolico) are now working in order to establish the criteria that would enable us to decide between variants in the critical edition of Maurolico's *Conics* we are preparing.

Maurolico's Reconstruction of the Fifth and Sixth Book of Appolonius's[1] *Conics*

Aldo Brigaglia

Quod si quis non credat, periculum faciat in quocumque voluerit theoremate vel problemate : cognoscet me non mentitum.

Maurolico was very proud of his capacity to reconstruct the lost works of classical Greek mathematics without knowing directly the original texts and in this quotation he is defying his reader to give him the lie. Without going deep in this problem (I refer for it to the other reports in this symposium) I may only add that Clagett's opinion[2] : " Maurolico's interest, unlike Commandino's, was not in establishing a philologically sound version or translation of the Archimedean texts that survived, but in presenting mathematically coherent texts that achieved Archimedes' objectives ", has been more and more firmly established in more recent years by many scholars (see for instance, Moscheo[3], Napolitani[4], Saito[5], Tassora[6], Gatto[7], Maierù[8]).

I will focus my attention on Maurolico's reconstruction of the (by then) lost books V and VI of Apollonius' *Conics* and I will not deal in any way with the

1. Apollonius de Perge, *Les Coniques*, translated and annoted by P. Ver Eecke, Paris, 1959.

2. M. Clagett, " Archimedes in the Middle Ages ", *The American philosophical Soc.*, III, Philadelphia, 1978.

3. R. Moscheo, *F. Maurolico tra rinascimento e scienza galileiana*, Messina, Soc. Messinese di Storia Patria, 1988 ; R. Moscheo, *Mecenatismo e scienza nella Sicilia del '500*, Messina, Soc. Messinese di Storia Patria, 1990.

4. P. Napolitani, " Maurolico e Commandino ", in P. Nastasi (ed.), *Il Meridione e le Scienze*, Univ. di Palermo, 1988, 281-316.

5. K. Saito, " Quelques observations sur l'édition des Coniques d'Apollonius de F. Maurolico ", *Boll. di Scienze Mat.*, XIV (1994), 239-258.

6. R. Tassora, *Il giovane Maurolico e lo studio delle sezioni coniche*, Tesi di Laurea, rel. P. Napolitani, Pisa, 1995.

7. R. Gatto, *Tra scienza e immaginazione. Le matematiche presso il collegio gesuitico napoletano (1522-1670)*, Firenze, 1994.

8. L. Maierù, *La teoria delle coniche nel cinquecento*, Sciascia, 1996.

interesting problem of Maurolico's sources[9] (see also Halley[10]). As far as I know, it is well documented only his knowledge of Memo's edition of the first four books of Apollonius' masterpiece and we have no evidence of his knowledge of Pappus' *Collections*[11]. So we may think of it as a completely original work by the Sicilian mathematician.

So the only hints Maurolico probably had in mind to build his *divinatio* were the two sketchy references given by Apollonius in his prefatory letter to Eudemus (I will always quote Apollonius by Maurolico's *Emendatio et Restitutio*)[12] : *Est enim quintus quidem de minimis & maximis, ut plurimum. Sextus de aequalibus & similibus coni sectionibus.*

Now I want to recall briefly the very definition of conic used mainly by Maurolico in his *Restitutio*. I refer to fig. 1 and I will only mention the hyperbole (it is very easy to shift to the definition of ellipse and parabola). I will also refer (as Maurolico does) only to the principal diameter, *diametris praecipuis, quae sunt axes* (*i.e.* the axe of the conic).

If we pose *ab* as the diameter of the hyperbole, *at* its parameter, *ah* as the abscissa of a point *z* in it and *hc* as the segment obtained in the figure as illustrated, the main *symptom* used by Maurolico is : $hz^2 = ah \cdot hc$ which is contained in proposition I. 21 of Apollonius, but which becomes in Maurolico's text, the main " plane " definition of a conic.

FIGURE 1

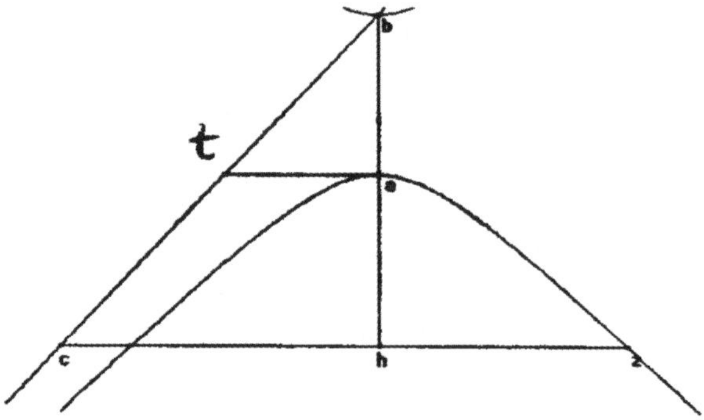

9. F. Amodeo, " Il trattato delle coniche di F. Maurolico ", *Bibliotheca math.* (4), 9 (1908), 123-138.

10. E. Halley, *Apollonii Pergaei conicorum libri octo et Sereni Antissensis de sectione cylindri libri duo*, Oxford, 1710.

11. Pappus d'Alexandrie, *La collection mathématique*, with an introduction and annoted by P. Ver Eecke, Paris, 1982.

12. F. Maurolico, *Emendatio et Restitutio Conicorum Apollonii Pergaei*, Messanae, 1654.

The other fundamental relation used by Maurolico is the very definition of parameter and of diameter of a conic. I briefly recall it.

If we call *abc* the triangle determined by a plane passing through the axe of a cone, and if we intersect the cone by another plane (α) and call *ak* the parallel to α driven in *abc* from *a* to *bc*, it is well known that the point *k* is in the interior of *bc* if the conic cut by α is a hyperbole, in the exterior if α it is an ellipse, coincides with *b* or *c* if it is a parabola. In any case (with the obvious exception of the parabola) the ratio between *ak* and the rectangle *bk · kc* is the same that the ratio between the diameter and the parameter of the conic.

So the problem to construct a conic similar to a given one (*i.e.* with a given ratio between diameter and parameter) amounts to find a point in the basis (or in its prolongation) of a given triangle such that the ratio between the square of the segment joining the vertex with this point and the rectangle of the two segments of the basis is also given.

FIGURE 2

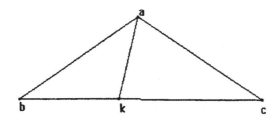

I now turn to a brief examination of some of the features of Maurolico's reconstruction of the v book of conics. We must immediately say that the *divinatio* of the Sicilian geometer is completely different from Apollonius' text. In fact the *maxima et minima* in the work of Apollonius are intended to find the minimal lines that may be driven from a point to a given conic (*i.e.* the normals to the conic) while Maurolico looks for the largest and the smallest among the conics with the same axe, the same vertex and the same diameter (in other words he is in some way interested to osculating conics). In what follows a conic will be said to be greater than another one if the second is entirely enclosed in the interior of the first one.

The Prop. 1 says that if two conics have equal name, vertex, axe and diameter, the two will be tangent in the vertex and the one with the greater parameter is greater than the other. The proof is direct using Maurolico's plane generation of the conics given above. I may limit myself to show the figure (fig. 3).

FIGURE 3

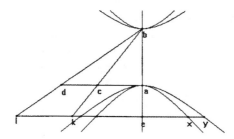

   In fact, calling *ac* and *ad* the two parameters, *ab* the diameter common to
the two hyperboles, and remembering that, for every abscissa *e*, the ordinate *ex*
is the proportional mean between *ae* and *ek*, while *ey* is the proportional mean
between *ec* and *el*, Maurolico's argument is self evident. (It is very easy to
modify slightly the proof for ellipse and parabola). A similar argument may be
used to compare two conics of the same type with the same parameter and
different diameters.

   Maurolico' s next aim is to compare conics of different type (always with
the same vertex, axe and diameter). I give, for instance, the case of an hyperbole and an ellipse. In any case they are tangent in the vertex. Moreover :

   1) If they have the same parameter, the ellipse is the smaller ; (Maurolico's
      V. 9)

   2) Among all the ellipses smaller than a given hyperbole, the greater is the
      one with the parameter equal to that of the hyperbole. (Maurolico's v.
      10). Moreover if an ellipse has the same diameter and a parameter greater
      than an hyperbole, it will meet the hyperbole exactly in two points more,
      besides the vertex.

   The proofs are again self evident if one remembers the planar construction
given above. I limit myself to give two figures (4-5) related to prop. 9 and
prop. 10 respectively.

FIGURES 4 AND 5

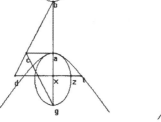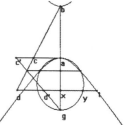

Proposition V. 28 is in some way recapitulatory. If different types of conics are put with the same vertex, axe and parameter they will be all tangent in the vertex, and, in decreasing order, the hyperbole, the parabola, the ellipse with the minor diameter in the axe, the circle, the ellipse with the major diameter in the axe. I show this in fig. 6.

FIGURE 6

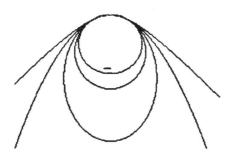

It is perhaps worth noting that this corresponds very well to Kepler's *system of conics*[13].

In the sixth book Maurolico is much nearer to Apollonius' original. In this case indeed it may be useful to compare the different approaches.

I begin with the very definition of equality and similarity of two conics. In Maurolico's words the two definitions are : *Similes conicae sectiones sunt, in quibus axium segmenta diametris proportionalia, ordinate ductas semper sibi proportionale suscipiunt.*

Aequales autem & similes conicae sectiones dicuntur in quibus axium segmenta aequalia ordinate ductas semper aequales suscipiunt.

Apollonius' definitions are instead (I quote from Toomer[14]) : " Conic sections which are called equal are those which can be fitted, one on another, so that the one does not exceed the other. Those which are said to be unequal are those for which that is not so.

And similar are such that, when ordinates are drawn in them to fall on the axes, the ratios of the ordinates to the lengths they cut off from the axes from the vertex of the section are equal to one another, while the ratios to each other

---

13. J. Kepler, *Paralipomena ad Vitellionem quibus astronomiae pars optica traditur*, Frankfurt, 1604.

14. G.J. Toomer (ed.), *Apollonius Conics Books V to VII. The Arabic translation of the lost greek original in the version of the Banu Musa*, New York, Springer, 1990, 264.

of the portions which the ordinates cut off from the axes are equal ratios. Sections which are dissimilar are those in which what we stated does not occur ".

It is worth noting above all the definition of equality : while Apollonius uses a classical Euclidean definition by superposition, Maurolico prefers to use a more " functional " definition : two conics are said to be equal if to each abscissa (*axium segmenta*) always corresponds (*semper suscipiunt*) an equal ordinate (*ordinate ductas*). Clearly the difference vanishes in the sequel. The definition chosen by Maurolico becomes a theorem in Apollonius.

I will now limit myself to sketch the main issue in this book (either in Apollonius as in Maurolico) : to cut a cone with a conic equal to a given one. The case of the parabola is the easier and I will not tract it.

Let us examine the case of the hyperbola. First of all Maurolico shows (prop. 3) that two hyperbolas are similar if and only if they have proportional parameter and diameter. From that he deduces that two hyperbolas cut in a cone by parallel planes are always similar (prop. 5).

Now he may consider the problem (prop. 10) to cut a cone with a plane in a hyperbola similar to a given one (and so in infinitely many, changing the plane with another parallel to the first one). If we remember what said in the introduction the problem amounts to this :

Given a triangle *abc*, to connect the vertex *a* with a point *d* in *bc* in such a way that the ratio between the square of *ad* and the rectangle *bd*, *cd* is given or, using Maurolico's notations the ratio between ad and the proportional mean between *bd*, *cd* (from now on *mp(bd,cd)*) is given (say *o:p*). The lemmas contained in prop. 7 and 8 solve completely this problem giving also its *determinatio* (*i.e.* the conditions for the problem to be solvable).

Let's begin from this *determinatio* (see fig. 7) :

If a triangle *abc* is given, among all the lines connecting a given vertex (say *a*) with the basis, calling *d* the point in which a line intersects the basis, the bisector of the angle in a has the least ratio between the mp(*bd*, *cd*)) and *ad*. Moreover, if two lines (say *ae* and *af*) are such that *eâd=fâd* the analogous ratios are equal. In other words we must prove that, in the said hypotheses

(1)

$$\forall e, \frac{ad}{mp(bd,cd)} < \frac{ae}{mp(eb,ec)}$$

(2)

$$\frac{ae}{mp(be,ce)} = \frac{af}{mp(bf,cf)}$$

FIGURE 7

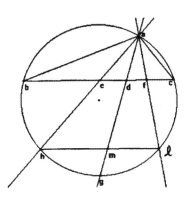

In fact we have (I use the symbols as given by figure 7) that *hl* is parallel to *bc* (because the arc *bh* is equal to the arc *cl* and so we have :

$$\frac{ad}{dm} = \frac{ae}{eh} > \frac{ad}{dg} \Rightarrow \frac{ae}{mp(ae,he)} > \frac{ad}{mp(ad,dg)}$$

moreover as it is well known we have : $mp(ae,he) = mp(be,ec)$ and $mp(ad,dg) = mp(bd,dc)$ and we get (1). The proof of (2) is trivial.

The proof given by Maurolico has an unmistakable apollonian flavour (see for example his proof of I. 54, and, for more information Saito[15], and gives us a clear insight of his mastery of classical methods.

Now we may turn to the construction above mentioned. Owing to the *determinatio* the given ratio (*o:p*) must be lesser than the ratio between the segments constructed from the bisector.

Maurolico's construction proceeds in this way : we draw the bisector of *â* and call it *ad* (look at fig. 7). After we take *r* as the third proportional after *o* and *p*. If we have *ad:dg = o:p* we are done. Otherwise take in *ag* a point *m* such that *am:mg = o:p* (it does exist owing to the *determinatio*). If from *m* we draw the parallel to *bc* and call its intersections with the circle circumscribed *e, f*. The two lines *af* and *ae* will be our solutions. The proof is now easy and I leave it to the reader[16].

15. K. Saito, " Compounded ratio in Euclid and Apollonius ", *Hist. Scientiarum*, 31 (1986), 25-59.

16. A. Brigaglia, " La ricostruzione dei libri V e VI delle Coniche da parte di F. Maurolico ", to appear in *Boll. di Storia delle Scienze mat.* (1998).

Now it is also easy to prove Maurolico's VI. 10 : to cut a given cone in a given hyperbola. First of all we must have that the ratio between the diameter and the *pm* between diameter and parameter (call it *o:p*) must not exceed the ratio between the bisector of the angle in the vertex of the triangle (call it *abc*) cut by the plane passing through the axe of the cone and the proportional mean between the segments cut in the basis.

If this is the case you may use the prop. VI. 8 to the triangle *abc* and find the points *e, f* such that satisfy the

$$\frac{ae}{mp(be,ce)} = \frac{af}{mp(bf,cf)} = \frac{o}{p}$$

and cut the cone with a plane parallel to *ae* or to *af* to get an hyperbola similar to the given one.

To complete these kinds of constructions we may cut a given cone with an hyperbola equal to a given one (Maurolico's proposition VI. 11). This construction amounts to choosing among the infinite hyperbolas similar to the given one, the hyperbola with the same diameter as the given. In turn this amounts to solve an easy problem of *neusis* : to insert in a given angle a segment of given length parallel to a given line.

It is worth noting that this problem is not solved by Apollonius. We may find its solution in the Pappus' lemmas added to the sixth book of the *Conics* (7.218). I may sketch Maurolico's construction in this way : let *abc* be the given angle, *ze* the given segment and *al* the direction of the given line. (See fig. 8). If we call *r* the point lying in *ba* such that *br* is the fourth proportional after *al, ze* and *ba*, the line through *r* parallel to *al* will solve our problem. In this way we may easily complete the construction of the hyperbola equal to the given one. Evidently the *determinatio* of the problem is the same of prop. VI. 11.

FIGURE 8

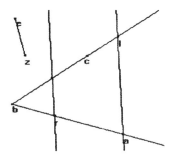

In my opinion it may be now interesting to compare briefly Maurolico's construction of the hyperbola equal to a given one with the analogous construction of Apollonius (prop. VI. 29). The two constructions are indeed almost identical, but it is surprising that Apollonius solves this problem only in a particular case (*i.e.* when we have a right cone). It is evident that in this case the problem is much easier.

First of all I want to note that this is a very strange choice by Apollonius, who always puts himself in the most general situation. Why did Apollonius only in this case choose restrictive hypotheses ? We could think to a copier's option, but it seems that all the extant copies of the 6th book of *Conics* are perfectly concord. This strange fact has been already noted by (Zeuthen[17]) and by (Toomer[18]) ; it seems however to me that neither has been aware of the extant general solution by Maurolico or have underlined the huge difference between the particular versus the general solution.

Toomer[19] says that " It is easy to see that his [Apollonius'] solutions in book VI can be extended to the oblique cone ". This is true only regarding to the way to cut the cone, but it is completely different regarding the *determinatio* of the problem. Indeed let us look for a moment the case of a right cone : the triangle *abc* will now be isosceles and so the bisector coincide with the altitude and the median and the *determinatio* is completely trivial.

When we have to construct an ellipse equal to a given one, the corresponding problem on triangles is similar to that regarding the construction of the hyperbola. In this case we must draw a segment connecting the vertex *a* to the basis protracted, as shown in fig. 9, and find a point *d* such that the ratio between *ad* and the proportional mean between the basis segments *db* and *dc* is a given one.

FIGURE 9

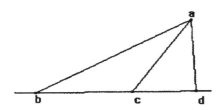

As in the former case we now have a cumbersome *determinatio* of the problem. For the case given in fig. 9 (of a triangle obtuse in *c*) I have drawn a

17. H.G. Zeuthen, *Die Lehre von den Kegelschnitten im Alterturm*, Copenhagen, 1886.

18. G.J. Toomer (ed.), *Apollonius Conics Books V to VII, op. cit.*

19. G.J. Toomer (ed.), *Apollonius Conics Books V to VII, op. cit.*, LVIII.

graph showing in ordinate the said ratio and in abscissa the position of point
d. The graph is given in fig. 10. The part between the two vertical asymptotes
represents the case of the hyperbola and has a minimum corresponding to the
bisector ; the left side is decreasing and always greater than one ; the right side
as a minimum lesser than one and after grows remaining always lesser than
one. The right and the left side correspond to the case of the ellipse. All this is
clearly and carefully shown by geometrical means by Maurolico. You may see
clearly the difference with the particular case studied by Apollonius comparing
fig. 10 with fig. 11 which shows the same graph in the case of an isosceles
triangle.

FIGURE 10

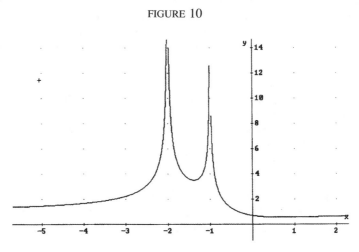

FIGURE 11

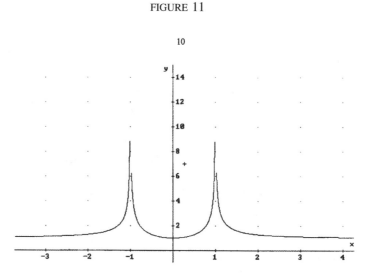

In conclusion I want to say only a few words. The *Emendatio et restitutio* by Francesco Maurolico is a stirringly original work, which contains some beautiful new results. Among them I want to cite the entire 5th book on the *systems of tangent* conics and part of the 6th book, particularly the construction of a conic equal to a given one in the case of the oblique cone. These are certainly some of the major results in the geometry of the first half of the 16th century. The work by Maurolico remained unpublished until the second half of the 17th century and the circulation of the manuscripts is not really clear (for information on this subject see Moscheo[20], 1988), so it is difficult to evaluate the influence of Maurolico's work on the scholars of his times. Claims (by Borelli and others) of Maurolico's direct influence on the work by Mydorge, Cavalieri, Kepler, Gregory of S. Vincent in my opinion are till now far from proven. This is a problem well worth to be further investigated.

20. R. Moscheo, *F. Maurolico tra rinascimento e scienza galileiana, op. cit.*

LA VERSION DE MAUROLICO DES *SPHÉRIQUES* DE MÉNÉLAOS ET SES SOURCES

Abdel-Kaddous TAHA, Pierre PINEL

PRÉSENTATION[1]

C'est en 1558 que parut à Messine la version des *Sphériques* de Ménélaos due à F. Maurolico[2]. Deux ans auparavant, il écrivait fièrement au vice-roi de Sicile, Juan de Vega : *Mihi olim libros curiose perquirenti oblata sunt exemplaria duo Menelai manu scripta in membranis : quorum utrumque mutilatum et mox male resartum fuerat (…). Neque author ipse, si revixisset, suum agnovisset opus. Hoc ego, antea diu desideratum, multo labore vigiliisque revisum restituere conatus compluribus tam jucundis, quam necessariis adauxi propositionibus*[3].

Maurolico nous dit donc s'être appuyé sur deux manuscrits, qu'il qualifie de " mutilés et mal réparés ", ou de " corrompus et mal corrigés " : ces deux interprétations, portant sur le contenant ou le contenu, sont en effet envisageables pour *mutilatum et male resartum*. Mais quels étaient ces deux manuscrits ? Les écrits de Maurolico ne contiennent aucune indication à ce sujet, et les historiens n'ont pu jusqu'à présent formuler que des conjectures.

---

1. Le lecteur intéressé par le sujet de cette communication en trouvera un exposé plus détaillé dans notre article intitulé " Sur les sources de la version de Maurolico des *Sphériques* de Ménélaos ", *Bollettino di Storia delle Scienze Matematiche*, vol. XVII, fasc. 2 (1997), 149-198.

2. *Cf. Biographical Dictionary of Mathematicians*, vol. III, New York, 1991, 1674-1678, ainsi que R. Moscheo, *F. Maurolico tra Rinascimento e scienza galileiana - Materiali e ricerche*, Messina, Società Messinese di Storia Patria, 1988.

3. C'est-à-dire : " A une époque où la recherche de livres m'occupait beaucoup, on m'a offert deux exemplaires des manuscrits de Ménélaos sur parchemin. Tous deux altérés, ils avaient été mal restaurés (…), à tel point que si l'auteur était revenu à la vie, il n'aurait pas reconnu son ouvrage. Mais moi, j'ai entrepris de reconstituer cette oeuvre que j'avais si longtemps cherchée, je lui ai consacré toutes mes peines et mes nuits, et je lui ai adjoint bon nombre de propositions aussi plaisantes qu'indispensables ". Cette lettre a été publiée dans le *Bullettino di bibliografia e di storia delle scienze matematiche e fisiche (Bullettino Boncompagni)* Tome IV (Janvier-Février 1876), n° IX, ainsi que dans l'ouvrage de G. Macrí, *F. Maurolico nella vita e negli scripti*, Messine, 1901, XLIX-LXXVI.

Pour éclaircir cette question, nous commencerons par comparer la traduction latine, fort répandue, réalisée par Gérard de Crémone au XIIᵉ siècle[4] avec la version de Maurolico ; nous en conclurons que, si Maurolico s'est certainement guidé sur Gérard de Crémone, il a utilisé un autre texte, qui ne pouvait être qu'un manuscrit arabo-musulman.

Après avoir évoqué la question de la connaissance de la langue arabe par Maurolico, nous chercherons à identifier ce deuxième texte. Nous écarterons tout d'abord la version la plus répandue, celle d'at-Tusi, ainsi que certaines autres versions arabes envisageables ; puis nous confronterons la version de Maurolico avec la version écrite au XIIIᵉ siècle par Jamal ad-Din Muhammad ibn Kamal ad-Din ibn al-'Adim 'Umar ibn Hibatallah ibn Muhammad ibn Abi Jarada tilmid (élève de) Muhammad ibn Wasil[5].

A cette fin, nous comparerons la structure et le contenu des versions de Jamal ad-Din et de Maurolico, et nous mettrons en évidence, à partir de quelques propositions représentatives, des ressemblances entre les versions de Maurolico et de Jamal ad-Din qui ne sauraient être dues au hasard.

Nous conclurons, malgré les quelques différences existantes entre ces deux versions, à l'utilisation de la version de Jamal ad-Din par Maurolico dans son travail de révision et de reconstitution des *Sphériques* de Ménélaos à partir de la version de Gérard de Crémone.

Cette conclusion n'est pas qu'anecdotique : elle identifie précisément un nouvel élément du treillis de transmission vers l'Occident de la Renaissance de l'héritage antique repris et étendu par les lettrés arabo-musulmans, et met en valeur l'importance de la Sicile comme point de contact des deux civilisations qui se disputaient la Méditerranée au XVIᵉ siècle.

LE PROBLÈME DES SOURCES DE LA VERSION DE MAUROLICO

Maurolico affirme en 1528 qu'il a démontré toutes les conclusions de Ménélaos sur la sphérique sans avoir jamais vu ses *Sphériques*[6] ; mais, pour pouvoir dire cela, il fallait bien qu'il ait connu le contenu de ces *Sphériques* et le nom de leur auteur, par exemple par l'intermédiaire d'un abrégé.

Dans son préambule (la *praefatio*) aux *Sphériques*, imprimées en 1558, Maurolico laisse entendre en revanche qu'il s'est servi d'un seul manuscrit. Il écrit en effet *Hos Menelai libellos cum ego in antiquis ex membrana codicibus*

---

4. *Cf.* P. Pinel et A.K. Taha, " Sur une version arabe anonyme précoce des *Sphériques* de Ménélaos conservée dans la traduction latine de G. de Crémone ", *5ᵉ Colloque Maghrébin sur l'Histoire des Mathématiques Arabes*, Tunis (1-3 Décembre 1994).

5. Que nous appellerons, au bénéfice de la simplicité, Jamal ad-Din dans la suite du présent travail.

6. " Omnes Menelai de Sphaericis conclusiones ostendi ; necdum Menelai Sphaerica vidi " — cité par M. Clagett, *Archimedes in the Middle Ages*, vol. III, Philadelphia, American Philosophical Society, 1964-1984, 771.

reperissem, conatus sum eos, quoniam corruptissimum erat exemplar, emendare ac restituere[7] ; il évoque donc un seul " exemplaire complètement corrompu ".

Enfin, la lettre d'août 1556 au vice-roi Juan de Vega, que nous avons citée dans notre introduction, indique formellement que Maurolico a travaillé à partir de deux sources manuscrites. Cela pourrait indiquer que Maurolico a rédigé son préambule bien avant août 1556, et n'a eu le deuxième manuscrit entre les mains que postérieurement ; et que, bien qu'il ait modifié le texte des *Sphériques* en fonction du contenu de ce deuxième manuscrit, il a négligé d'en corriger le préambule.

Nous allons maintenant essayer d'identifier les deux sources en question, qui ne pouvaient être que latines ou arabes, les originaux grecs des *Sphériques* de Ménélaos étaient perdus depuis longtemps à l'époque de Maurolico.

MAUROLICO ET GÉRARD DE CRÉMONE

Nous savons que Maurolico connaissait les écrits de Gérard de Crémone. Il fait allusion en effet à celui-ci, à plusieurs reprises, dans ses lettres. Il en parle par exemple, de façon peu élogieuse, dans sa lettre du 4 mai 1536 au cardinal Pietro Bembo[8] : *Quod si Ioannes a regio monte aut eius praeceptor Georgius Peurbachius, uterque mathematicus consummatissimus, sicut planetarum theorias, sic Spherae rudimenta nobis edidisse[nt], iam pridem cum Gerardo Cremonensi una Ioannes hic de Sacro Bosco rudis astronomus exsibilatus et explosus fuisset*[9].

Björnbo, qui a étudié de façon approfondie les versions latines des *Sphériques* de Ménélaos, estime cependant que Maurolico ne s'est pas servi de la version de Gérard de Crémone[10], au vu des différences entre celle-ci et le texte

---

7. " Ayant trouvé ces écrits de Ménélaos parmi de vieux livres en parchemin, je me suis efforcé, comme l'exemplaire était complètement corrompu, de les corriger et de les reconstituer ".

8. *Cf.* Moscheo, F. *Maurolico tra Rinascimento e scienza galileiana - Materiali e ricerche, op. cit.*, 273.

9. " Et si J. Regiomontanus ou son maître G. Peurbach, tous les deux des mathématiciens de très grande valeur, nous avaient exposé les rudiments de la sphérique comme ils ont exposé la théorie des planètes, il y a longtemps déjà que cet astronome ignorant qu'est J. Sacrobosco aurait été chassé sous les sifflets et les huées avec G. de Crémone ". On rencontre des remarques analogues dans la préface de Maurolico à son édition de 1567 de la Sphère de Sacrobosco (*cf.* Moscheo, F. *Maurolico tra Rinascimento e scienza galileiana - Materiali e ricerche, op. cit.*, 197).

10. Voir A.A. Björnbo, " Studien über Menelaos' Sphärik. Beiträge zur Geschichte der Sphärik und Trigonometrie der Griechen " [Études sur la Sphérique de Ménélaos. Contributions à l'histoire de la sphérique et de la trigonométrie des Grecs], *Abhandlungen zur Geschichte der mathematischen Wissenschaften mit Einschluss ihrer Anwendungen*, vol. 14 (Leipzig, 1902). Il écrit à ce sujet " la différence [entre l'édition de Maurolico et la traduction de G. de Crémone] nous montre que l'hypothèse reçue jusqu'ici, selon laquelle Maurolico se serait servi de cette traduction, est fausse. Le plus vraisemblable est qu'il a utilisé des manuscrits arabes (…) " (A.A. Björnbo, " Studien über Menelaos' Sphärik. Beiträge zur Geschichte der Sphärik und Trigonometrie der Griechen ", *op. cit.*, 19-20).

de Maurolico. Comme Montcula[11], il suppose que Maurolico s'est servi de manuscrits arabes.

Il est cependant probable que Maurolico a eu en mains la version des *Sphériques* de Gérard de Crémone, qui figurait couramment dans les codex traitant de la sphérique — ou au moins des parties importantes de ce texte. Nous allons essayer de corroborer cette hypothèse en examinant les analogies entre la version de Gérard de Crémone et celle de Maurolico ; puis nous déduirons des dissemblances entre ces versions que la version de Gérard de Crémone n'a pas été la source unique de Maurolico.

A. Points communs essentiels entre les versions de Maurolico et de Gérard de Crémone

1. Maurolico utilise constamment, jusqu'à la proposition III-9[12], l'expression par laquelle Gérard de Crémone désigne le triangle sphérique, *triangulus ex arcubus circulorum magnorum super superficiem spherae*, alors que les Arabes utilisent " triangle " (*muthallath*, littéralement : objet composé de trois éléments) sauf Abu Nasr, qui parle de " figure à trois côtés " (*shakl dhu thalatha adla'*).

2. Les lettres utilisées dans les figures par Maurolico sont identiques à celles de Gérard de Crémone (sauf aux théorèmes II-21 et II-22[13], et sauf bien sûr dans les cas où la démonstration est propre à Maurolico et où il utilise donc une figure différente de celle de Gérard de Crémone). Ces lettres sont par contre souvent différentes de celles utilisées par les auteurs arabes : nous n'avons pu établir aucun système de translittération cohérent entre un auteur arabe et Maurolico, les problèmes se posant particulièrement pour le *'ain*, le *sad* et le *sin*.

Or, il est peu probable que Maurolico ait pu réinventer *ex nihilo* certaines correspondances entre lettres arabes et lettres latines utilisées par Gérard de Crémone ; celui-ci transcrit, par exemple, le *ha* en *p* et le *tha* en *y*.

Enfin, on peut être sûr que Maurolico a suivi un modèle pour la désignation des points des figures ; lorsqu'il introduit une figure qui lui est personnelle, il nomme les points de cette dernière dans l'ordre de l'alphabet (grec dans sa version des *Sphériques* de Ménélaos, latin dans ses propres *Sphériques*).

3. On peut encore constater des concordances significatives dans les énoncés de certaines propositions du livre I[14] (elles sont particulièrement nettes, par exemple, aux propositions I-9 et I-23, qui correspondent aux propositions I-6 et

---

11. J.E. Montcula, *Histoire des Mathématiques*, vol. I, Paris, 1758, 563.

12. Par la suite, sauf dans la proposition III-11 (bis), il utilise l'expression *triangulus sphaeralis*.

13. Dans ces théorèmes, Maurolico remplace par un *t* le *y* de G. de Crémone.

14. *Cf.* la thèse de J.P. Sutto : *F. Maurolico, mathématicien italien de la Renaissance (1494-1575)*, Paris 7, Université D. Diderot, juin 1998.

I-14c chez Gérard de Crémone), ainsi que l'utilisation par Maurolico de certaines figures spécifiques à Gérard de Crémone (comme à la proposition I-23 déjà évoquée). Ainsi, les deux auteurs utilisent une seule figure pour les propositions I-24 et I-25 (I-15 et I-16 chez Gérard de Crémone), alors que les autres versions se servent de deux figures distinctes.

B. Différences entre ces versions

Il est cependant douteux que l'édition de Gérard de Crémone ait été la source unique de Maurolico, que celui-ci se serait contenté de corriger et de compléter selon son inspiration.

On peut relever en effet des différences importantes entre leurs textes. Nous allons examiner de plus près quelques théorèmes significatifs, répartis dans l'ensemble de l'ouvrage :

1. Erreur chez Gérard de Crémone, absente chez Maurolico :

Dans la proposition I-22 (I-13 dans la numérotation de Gérard de Crémone), Maurolico montre l'égalité de deux triangles sphériques ABG, DEZ tels que les angles A et D soient égaux, que AG = DZ, GB = ZE, et que les angles restants angle B et angle E soient différents de deux droits.

La condition sur les angles B et E peut être interprétée de façons différentes ; le texte grec, conservé chez Théon[15], dit en effet : Ἐὰν δύο τρίπλευρα...χη,...τὰς δ λοιπὰς γωνίας ἅμα δυσὶν ὀρθαῖς μὴ ἴσας... et peut être compris comme signifiant que chacun des angles doit être différent d'un droit, ou comme signifiant que leur somme doit être différente de deux droits.

Chez Gérard de Crémone, c'est la première interprétation (fausse) qui est retenue ; chez certains auteurs arabes, on trouve la deuxième interprétation (exacte) ; chez Maurolico, on trouve une troisième formulation : les angles B et E doivent être simultanément aigus ou simultanément obtus (*aut ambo acuti, aut ambo obtusi*, écrit Maurolico), ce qui introduit une hypothèse plus forte que celle qui est nécessaire et suffisante[16].

2. Hypothèse inutile chez Maurolico, absente chez Gérard de Crémone :

Dans la proposition II-7 (I-38 dans la numérotation de Gérard de Crémone), Maurolico combine inutilement deux hypothèses en écrivant : *si fuerit ... duo arcus ... minus semicirculo, et maior eorum (cum inequales sunt) non maior quadrante*, alors que Gérard de Crémone suppose seulement que deux des arcs sont chacun inférieur à un droit (*si fuerit ... unusquisque duorum arcuum ... minor quarta circuli*).

---

15. Cité par A.A. Björnbo, " Studien über Menelaos' Sphärik. Beiträge zur Geschichte der Sphärik und Trigonometrie der Griechen ", *op. cit.*, 22.

16. Même si, combinée avec les autres hypothèses, cette formulation se ramène à la condition $B + E \neq \pi$.

3. Variantes présentes chez un auteur, absentes chez l'autre :

La proposition II-8 est une variante de la proposition II-7, dans laquelle Maurolico prend le point D sur la base AG, au lieu de le prendre à l'intérieur du triangle ABG. Cette variante n'existe pas dans la version de Gérard de Crémone.

Inversement, on considère dans la proposition III-6 (III-4 dans la numérotation de Gérard de Crémone) deux triangles sphériques ABG, DEZ dans lesquels on abaisse des perpendiculaires vers les bases AG et DZ. Ces perpendiculaires peuvent tomber à l'extérieur ou à l'intérieur des arcs AG et DZ. Gérard de Crémone dessine les deux figures possibles, alors que Maurolico ne donne qu'une seule figure. Ceci est d'autant plus étonnant que son habitude est plutôt de distinguer les cas de figure possibles et de tracer les dessins correspondants.

C. Conclusion

Nous avons constaté des ressemblances fondamentales entre la version de Maurolico et celle de Gérard de Crémone, portant aussi bien sur le vocabulaire utilisé pour désigner le triangle sphérique que sur le lettrage ou le tracé des figures ; nous avons également pu rapprocher les énoncés ou les figures de certaines propositions.

Il est difficile, dans ces conditions, d'écarter l'hypothèse que Maurolico se soit appuyé sur la version de Gérard de Crémone, d'autant plus que celle-ci lui était très certainement accessible.

En revanche, nous avons présenté quatre propositions parmi d'autres, issues des trois livres des *Sphériques*, où le contenu du texte de Maurolico s'écarte fortement de celui de Gérard de Crémone. Il existe de plus chez Maurolico quantité de propositions absentes de la version de Gérard de Crémone.

Ce dernier n'a donc pas pu être la source unique de la version de Maurolico.

MAUROLICO ET LES AUTEURS ARABES

La deuxième source de Maurolico, compte tenu de l'histoire de la transmission du texte, n'a pu être qu'une version arabo-musulmane, que nous allons maintenant chercher à identifier.

A. Les travaux arabo-musulmans sur les *Sphériques* de Ménélaos

Les *Sphériques* ont fait partie des premiers ouvrages scientifiques grecs traduits en arabe[17]. La traduction *princeps* pourrait avoir été établie vers l'an 200 de l'Hégire, soit en 815-816 après J.-C., peut-être à partir d'une version syriaque. D'autres versions et traductions, pour la plupart perdues, ont vu le jour aux IXe et Xe siècles, dues à de grands auteurs comme al-Mahani, Ishaq ibn Hunayn, Abu Uthman ad-Dimashqi, Yuhanna ibn Yusuf ou al-Harawi.

On connaît encore une autre version, anonyme, conservée dans les traductions latine de Gérard de Crémone (XIIe siècle) et hébraïque de Jacob ben

Makhir (XIIIe siècle), ainsi que dans des fragments de l'*Istikmal*, ouvrage mathématique écrit par le prince de Saragosse Ibn Hud à la fin du XIe siècle[18].

Ensuite sont venues la version commentée écrite en 1007-1008 par le prince Abu Nasr ibn `Iraq, puis celle qui a été réalisée en 1265 par le célèbre astronome et mathématicien at-Tusi ; cette dernière, qui a été la première édition explicitement critique des *Sphériques* de Ménélaos et a eu un grand retentissement[19], est conservée dans de nombreux manuscrits ; elle a été reprise par un collaborateur d'at-Tusi nommé al-Maghribi.

La dernière version écrite avant l'époque de Maurolico est due à Jamal ad-Din, qui l'a réalisée avant 1300 ; c'est la deuxième édition critique arabo-musulmane des *Sphériques*.

B. Les auteurs arabes pouvant être écartés

Le premier ouvrage arabe envisageable comme source pour Maurolico est bien sûr l'excellente version d'at-Tusi, qui était la plus répandue en Orient.

Or, à la proposition I-22 évoquée précédemment, l'hypothèse trop forte considérée par Maurolico avait été explicitement critiquée par at-Tusi, qui se contente de supposer à juste titre que la somme des angles B et E est inférieure à deux droits. Une erreur de même type a été commise par Maurolico à la proposition I-28, où il suffit, comme le précise at-Tusi, de supposer que la somme des angles B et E est supérieure ou égale à deux droits, alors que Maurolico suppose chacun de ces deux angles supérieur ou égal à un droit. Enfin, Maurolico utilise à la proposition II-7 une hypothèse inutile, non mentionnée par at-Tusi. Ces trois arguments suffisent à écarter la version d'at-Tusi en tant que source utilisée par Maurolico.

Les mêmes arguments s'appliquent à la version d'Abu Nasr. La source arabe cherchée ne peut pas non plus être al-Maghribi, chez qui les démonstrations sont réduites à leur plus simple expression.

Compte tenu de la parenté existant entre les versions de Gérard de Crémone et d'al-Harawi, les différences entre les versions d'al-Harawi et de Maurolico

17. Voir M. Krause, " Die Sphärik von Menelaos aus Alexandrien in der Verbesserung von Abu Nasr Mansur b. 'Ali b. 'Iraq, mit Untersuchungen zur Geschichte des Textes bei den islamischen Mathematikern " [La Sphérique de Ménélaos d'Alexandrie dans la version améliorée par Abu Nasr Mansur ben 'Ali ben 'Iraq, avec des recherches pour contribuer à l'histoire du texte chez les mathématiciens arabes], *Abhandlungen der Gesellschaft der Wissenschaften zu Göttingen*, Philologisch-historische Klasse, 3e série, n° 17 (Berlin, 1936) ; A.K. Taha, " Les textes arabes des Sphériques de Ménélaos ", *4e Colloque Maghrébin sur l'Histoire des Mathématiques Arabes* (Fès, 2-4 Décembre 1992) ; voir aussi le *Dictionary of Scientific Biography*, vol. IX, 296-302.

18. Voir J.P. Hogendijk, " Which version of Menelaus' Spherics was used by Al-Mu'taman ibn Hud in his Istikmal ? ", in M. Folkerts (éd.), *Mathematische Probleme im Mittelalter - Der lateinische und arabische Sprachbereich* [Problèmes mathématiques au Moyen Age - le domaine linguistique latin et arabe], Wiesbaden, Harrassowitz Verlag, 1996, 17-44.

19. Voir, par exemple, G.E. Yusupova, " Kommentarii k " Sferike " Menelaja at-Tusi i al Yazdi " [Les commentaires sur la Sphérique de Ménélaos par at-Tusi et al-Yazdi], *Izvestija Akademii Nauk UZSSR, Serija Fizicheskikh-Matematicheskikh Nauk*, n° 6 (1990), 40-43 et 80.

sont moins tranchées. Il existe cependant des divergences significatives entre ces dernières : par exemple, al-Harawi démontre la deuxième partie du théorème I-37[20] grâce à une démonstration par l'absurde qui lui est propre et qui ne se retrouve pas chez Maurolico.

C. La version de Jamal ad-Din

Nous avons évoqué plus haut la version de Jamal ad-Din. Cet écrit a été peu commenté dans le passé[21], et semble peu connu des historiens des sciences[22]. Il est conservé, dans un unique manuscrit, à la bibliothèque de Manisa en Turquie sous la référence Genel 1706/1, ff. 34a-105b, sous le double titre " Commentaires sur les *Sphériques* de Ménélaos - Livres de Ménélaos sur les figures sphériques "[23].

1. L'auteur

Nous ne connaissons ni sa date de naissance ni sa date de mort ; la seule chose que nous puissions dire est qu'il a vécu au XIIIe siècle, car le manuscrit de son traité des *Sphériques* de Ménélaos n'a pu être réalisé qu'entre 1265 et 1299[24].

Il se pourrait qu'il s'agisse du même personnage que Muhammad ibn 'Umar ibn Ahmad Hibatallah ibn Abi Jarada, qui a commenté en 1292 le livre de Thabit ibn Qurra sur la section du cylindre[25], et qu'il ait été l'élève de Jamal ad-Din Abu Abdallah Muhammad ibn Salim ibn Wasil[26] (1207/08-1298), juriste, philosophe, mathématicien et astronome, envoyé en 1261 en ambassade auprès du roi Manfred de Sicile par le sultan d'Égypte Baybars, et qui a eu Abulfida comme élève ; ceci fournirait une hypothèse intéressante quant à la présence en Sicile de la version des *Sphériques* dont nous traitons.

20. Qui correspond au théorème I-44 chez Maurolico.

21. Les seuls commentaires connus sur le traité de J. ad-Din sont dus à M. ibn Baqir ibn Zain al-'Abidin al-Yazdi (mort en 1637). Celui-ci, qui a commenté les versions de Tusi et de Jamal ad-Din, considère les démonstrations de ce dernier comme les plus parfaites. La version d'al-Yazdi est conservée à la bibliothèque Saltykov-Chtchédrine, collection Khanikov, n° 144 de St Pétersbourg et à la bibliothèque privée Mu'tamid de Téhéran (cf. G.E. Yusupova, " Kommentarii k " Sferike " Menelaja at-Tusi i al Yazdi ", *op. cit.*, et Sezgin, *Geschichte des Arabischen Schrifttums*, vol. V, Leiden, 1974, 163).

22. A notre connaissance, cette version, qui est simplement signalée par Sezgin, *Geschichte des Arabischen Schrifttums*, *op. cit.*, n'a jamais fait l'objet d'une étude publiée.

23. Les pages 34b-48b contiennent un livre de 49 propositions sur les proportions dans le plan et sur la sphère dues à J. ad-Din, qu'il considère comme préalables à l'exposition des *Sphériques* de Ménélaos.

24. Sa rédaction est en effet postérieure à 1265, car J. ad-Din reprend certains commentaires d'at-Tusi (sans citer ce dernier) ; or, at-Tusi a terminé la rédaction de sa version en 1265. D'autre part, le manuscrit de Manisa a été recopié en l'an 699 de l'Hégire, soit en 1299 de l'ère chrétienne. Le traité a donc été rédigé entre 1265 et 1299.

25. *Cf.* Suter, *Die Mathematiker und Astronomen der Araber und ihre Werke* [Les mathématiciens et les astronomes des Arabes et leurs travaux], Leipzig, Teubner, 1900, 158 (n° 385).

26. Suter, *Die Mathematiker und Astronomen der Araber und ihre Werke*, *op. cit.*, 157 (n° 380).

2. *Le traité*

La version des *Sphériques* de Jamal ad-Din est divisée en 4 livres, à la place des 2 ou 3 livres qu'on trouve dans toutes les autres versions. Les livres I et II correspondent aux livres I et II d'at-Tusi ; le livre III contient les propositions III-1 à III-14 d'at-Tusi, et le livre IV ses propositions III-15 à III-25.

Le contenu de la version de Jamal ad-Din des *Sphériques* est globalement conforme à celui de la version " noire " utilisée par at-Tusi. Ses caractéristiques sont les suivantes :

1. Presque toutes les propositions sont agrémentées de commentaires, qui contiennent soit des éclaircissements, soit des démonstrations alternatives, soit des discussions sur les hypothèses, soit l'introduction de cas non traités par Ménélaos. La distinction entre le texte et les commentaires est clairement opérée par Jamal ad-Din.

2. Par contre, il n'indique pas les applications astronomiques de certains théorèmes, alors qu'Abu Nasr et at-Tusi le font.

3. Il est enfin à remarquer que, dans ses démonstrations, Jamal ad-Din utilise la corde de l'arc double[27], comme dans les versions les plus anciennes, alors qu'il était habituel depuis Abu Nasr d'employer le sinus à la place de cette grandeur.

On peut retrouver les traces d'au moins cinq versions des *Sphériques* dans le texte de Jamal ad-Din : la version que at-Tusi appelle la version " noire ", celle qu'il appelle la version " rouge ", la version d'al-Harawi, la version d'at-Tusi, et la version d'al-Maghribi[28].

D. Comparaison des versions de Maurolico et Jamal ad-Din

1. *Similitudes dans la structure et le contenu des ouvrages*

Maurolico divise apparemment les *Sphériques* de manière classique, en trois livres. Mais, si les deux premiers livres de son édition correspondent à ce qu'on rencontre chez les autres auteurs, son troisième livre s'arrête à la proposition III-13 (dans la numérotation d'at-Tusi), alors que les autres propositions du livre III forment le noyau de ses propres *Sphériques* ; on est en droit de considérer que sa version des *Sphériques* de Ménélaos et ses propres *Sphériques* forment en fait un tout en quatre parties.

Cette structuration tout à fait particulière correspond très précisément à la division des *Sphériques* de Ménélaos en quatre livres par Jamal ad-Din ; celui-ci opère la coupure entre le troisième et le quatrième livre au même endroit que

---

27. Nous savons cependant que J. ad-Din connaissait le sinus, puisqu'il le définit dans la proposition 36 de son livre sur les proportions évoqué à la note 23.

28. J. ad-Din mentionne également A. Nasr à propos du théorème III-5 ; il se peut cependant qu'il n'ait pas disposé de sa version des *Sphériques*, mais seulement de son commentaire sur ce théorème.

Maurolico, après la proposition III-13 ; aucun autre auteur n'a structuré les *Sphériques* en quatre livres.

D'autre part, nous avons comparé le contenu mathématique des versions de Maurolico, de Jamal ad-Din et d'at-Tusi, cette dernière résumant en quelque sorte toutes les versions arabo-musulmanes antérieures. Cette comparaison a fait apparaître l'existence de 21 propositions communes à Maurolico et à Jamal ad-Din, et à eux seuls. Il s'agit en général de prémisses ou de commentaires associés par Jamal ad-Din à des propositions du livre II, qui sont incluses par Maurolico en tant que propositions dans son texte[29].

2. *Proximité des démonstrations et de la démarche*

Nous avons déjà évoqué les hypothèses des propositions I-22, I-28, II-7 et II-8, qui particularisent la version de Maurolico par rapport aux versions de Gérard de Crémone et d'at-Tusi. A chacune de ces propositions, les hypothèses de Jamal ad-Din sont les mêmes que celles de Maurolico (trop fortes aux théorèmes I-22 et I-8, ou inutiles aux propositions II-7 et II-8).

Mais les points communs à ces deux dernières versions ne se limitent pas là. On trouvera ci-dessous plusieurs exemples particulièrement significatifs, comprenant :

- deux démonstrations où Maurolico suit un chemin emprunté par le seul Jamal ad-Din avant lui,

- six groupes cohérents de cas particuliers examinés par Jamal ad-Din et Maurolico, et par eux seuls,

- et un problème traité dans sa généralité seulement par nos deux auteurs.

a) *Des démonstrations simplifiées* (*propositions II-3 et II-41*)

- Dans la proposition II-3, il s'agit de démontrer qu'un arc tracé sous certaines conditions coupe obligatoirement un des côtés du triangle.

Dans la démonstration " classique " utilisée par tous les auteurs antérieurs à Jamal ad-Din, on prend un point D à l'intérieur du triangle, puis on trace les arcs BDE et DZ tels que les angles DZG et A soient égaux.

Jamal ad-Din reprend cette démonstration, mais en donne également une autre, plus simple, où on prend un point E sur la base, puis on trace les arcs BE et ED tels que les angles DEG et A soient égaux.

La démonstration de Maurolico suit le même cheminement que la démonstration alternative donnée par le seul Jamal ad-Din : il prend un point D sur la base, puis trace les arcs BD et DZ tels que les angles EDG et A soient égaux.

---

29. Il s'agit des propositions I-44 (lemme), II-2, II-3, II-12, II-15, II-18, II-23 et 24, II-30 et 31, II-34 à 36, II-38 à 40, II-41b et f, et II-43 à 45.

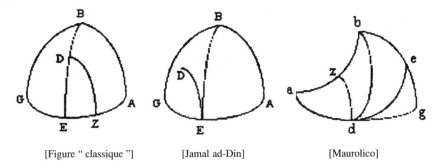

[Figure " classique "]   [Jamal ad-Din]   [Maurolico]

- Dans la proposition II-41, on suppose l'égalité de certains arcs découpés sur la base d'un triangle, et il s'agit de démontrer que les angles correspondants sont inégaux.

La démonstration " classique " de cette proposition repose sur la construction du triangle auxiliaire ANL et sur l'utilisation des théorèmes I-23 et II-7.

Par contre, Maurolico et Jamal ad-Din (ce dernier dans une démonstration alternative) se contentent de tracer un arc (CX avec les notations de Maurolico) et d'utiliser la proposition II-17. Ce sont les seuls à opérer ainsi.

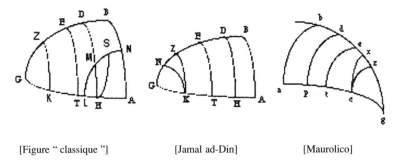

[Figure " classique "]   [Jamal ad-Din]   [Maurolico]

b) Six groupes de cas particuliers spécifiques (relatifs aux propositions II-9, II-10, II-21, II-29, II-33 et II-37)

Il s'agit par exemple, dans la proposition II-9, de tirer les conséquences du découpage de deux arcs BD et EZ sur le côté BG du triangle.

Maurolico et Jamal ad-Din traitent tous les deux, en plus du cas général, trois cas particuliers (D confondu avec E, Z confondu avec G, et la combinaison des deux cas précédents), et ils sont les seuls auteurs à le faire[30] ; ceci est d'autant plus remarquable que leur étude se borne là, et qu'ils n'étudient pas

---

30. Seul al-Maghribi évoque de façon parcellaire certains de ces cas particuliers, et il le fait sans donner de démonstration.

les autres cas envisageables (ceux où le début du premier arc découpé n'est pas confondu avec B).

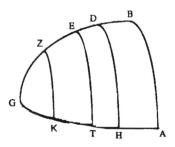

Cas traités					Maurolico	J.ad-Din
B	D	E	Z	G	II-9	II-6
B	D	E		G	II-11	II-6, com. 2
B		D		G	II-13	II-6, com. 3
B	D		E	G	II-15	II-6, com. 1
Cas non traités						

On retrouve exactement la même situation pour les découpages correspondant aux propositions II-10, II-21, II-29, II-33 et II-37.

c) Un problème traité dans sa généralité (propositions II-42 à II-45)

On considère, dans cet ensemble de propositions, deux arcs de grands cercles AB et GD qui rencontrent quatre grands cercles soit tous issus du pôle de AB ou de GD, soit tous tangents à un petit cercle parallèle à AB ou à GD.

Maurolico et Jamal ad-Din sont les seuls auteurs à avoir ainsi traité dans sa généralité le problème posé, les autres auteurs se contentant de traiter le cas des cercles issus du pôle de AB ou de GD. Ceci est d'autant plus significatif que les auteurs suivent le même plan, utilisent les mêmes théorèmes dans leurs démonstrations, indiquent les mêmes variantes et terminent tous deux en reprenant sous une formulation générale les résultats obtenus. Il est difficile de supposer que ces ressemblances sont le fait du hasard !

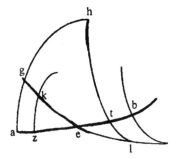

3. *Sur les différences entre les versions de Jamal ad-Din et de Maurolico*

Ces versions présentent un certain nombre de différences, dans l'ordre des propositions ou dans leur groupement ; certaines propositions sont propres à un auteur, et ne se retrouvent pas chez l'autre. Les formulations employées par Maurolico s'écartent souvent de celles de Jamal ad-Din ; par exemple, Maurolico utilise fréquemment le raisonnement par l'absurde, alors que Jamal ad-Din le fait rarement, seulement dans des démonstrations alternatives, et en indiquant que ceci s'écarte de Ménélaos.

Nous considérons qu'il s'agit là de différences somme toute secondaires, qui ne remettent pas en cause notre hypothèse.

En effet, Maurolico ne cherchait pas à faire oeuvre historique ou critique, mais à reconstituer et à présenter le plus clairement possible une théorie qui lui paraissait très importante. Il est donc normal qu'il ait amplement retravaillé ses sources.

D'autre part, il ne faut pas oublier qu'il ne disposait que d'" exemplaires altérés et mal restaurés ". Ceci peut expliquer l'absence de plusieurs théorèmes des *Sphériques*, situés à la fin du deuxième livre : les pages correspondantes manquaient peut-être sur les deux manuscrits dont s'est servi Maurolico.

E. Conclusion

Nous avons examiné tout d'abord la structure des versions de Jamal ad-Din et de Maurolico, et constaté que tous deux découpaient en réalité leurs *Sphériques* en quatre livres, et avec la même césure. Puis, nous avons constaté la proximité du contenu de ces versions, dans lesquelles apparaissent 21 théorèmes absents des autres versions. Nous avons trouvé ensuite dans la formulation des hypothèses de plusieurs théorèmes des erreurs communes à Jamal ad-Din et à Maurolico, et à eux seuls ; enfin, plusieurs démonstrations spécifiques à ces deux auteurs, et à eux seuls, ont été relevées.

L'ensemble de ces éléments, que ne remettent pas en question les différences perceptibles entre les deux versions examinées, nous permet d'affirmer, avec un haut degré de vraisemblance, que c'est la version de Jamal ad-Din qui a été la deuxième source des *Sphériques* de Maurolico.

CONCLUSION GÉNÉRALE

Nous espérons avoir éclairci la question posée dans l'introduction relativement aux sources utilisées par Maurolico pour écrire sa version des *Sphériques* de Ménélaos, et avoir prouvé qu'il est vraisemblable que, s'il s'est appuyé, comme on pouvait s'y attendre, sur l'édition latine de Gérard de Crémone, fort répandue au Moyen Age, il s'est servi pour améliorer son texte d'un ouvrage en langue arabe : la version de Jamal ad-Din (ou une version liée à celle-ci), que le hasard a mis entre les mains de Maurolico.

Malgré tout, des différences importantes subsistent entre Maurolico et Gérard de Crémone, ainsi qu'entre Maurolico et Jamal ad-Din. Maurolico ne s'est pas contenté, loin de là, de corriger et de compléter Gérard de Crémone à partir de Jamal ad-Din.

Il a imprimé puissamment sa marque aux *Sphériques*, en particulier dans le livre III, en réorganisant des théorèmes et en ajoutant des résultats qui lui sont propres[31] ; il a bouleversé plusieurs démonstrations, en utilisant par exemple le raisonnement par l'absurde ou en passant par une démonstration sur le plan pour prouver projectivement une propriété sphérique : il s'agit bien d'un travail *ex traditione Maurolyci* " selon l'édition de Maurolico ", comme l'indique la page de titre de l'édition de 1558.

---

31. Il est d'ailleurs possible qu'il ait utilisé, pour établir ces résultats, d'autres sources (*Traité du Quadrilatère* d'at-Tusi ou *Traité des Transversales* de Thabit ibn Qurra, par exemple).

LES ARITHMÉTIQUES DE FRANCESCO MAUROLICO

Jean-Pierre SUTTO

Il est possible de distinguer trois types d'arithmétiques dans l'oeuvre de Francesco Maurolico : l'arithmétique euclidienne, l'arithmétique néo-pythagoricienne et l'arithmétique en tant qu'instrument de ce que Maurolico appelle la quantité générale.

Sur la première, l'effort du sicilien porte sur la clarification et la réduction des démonstrations des éditions de Zamberti et Campanus. Pour la deuxième il s'agit d'une monumentale extension aux nombres figurés, présents à la Renaissance principalement dans les éditions de Boèce et de Jordanus. Le caractère de figure disparaît presque totalement, au profit d'une vision numérique. Enfin le troisième type est une théorie de la quantité générale, sorte de lieu commun entre quantité discrète et quantité continue.

Les résultats sur les opérations, la rationalité et la commensurabilité, sont communs aux lignes et aux nombres et peuvent être généralisés à la quantité générale. Maurolico cherche à construire un cadre théorique aux arithmétiques pratiques des nombres irrationnels qui fleurissent à la Renaissance, mais il se place surtout comme un élément incontournable dans l'histoire de la *Mathesis universalis* pour l'époque classique.

Un autre type d'arithmétique pourrait éventuellement avoir sa place dans ce recensement : l'arithmétique pratique. Il ne nous en reste que quelques traces, difficile à évaluer. Nous nous appuierons donc principalement sur l'édition posthume vénitienne de 1575, dont il existe un manuscrit non autographe et de nombreuses notes et fragments divers dans les manuscrits autographes et pour la plupart inédits de Maurolico. Les textes des arithmétiques de l'imprimé sont datés sans ambiguïté de 1557[1].

---

1. F. Maurolico, *Arithmeticorum libri duo*, Venise, 1575.

L'ARITHMÉTIQUE EUCLIDIENNE

Nous passerons très vite, à tort sans doute, sur l'arithmétique euclidienne de Maurolico. Le Sicilien écrit en 1534 une édition des *Éléments*, dont la partie strictement géométrique est perdue et dont le reste demeure sous forme manuscrite, auquel il manque encore les énoncés. Il écrit dans les dernières années de sa carrière un fort intéressant compendium des *Éléments*, inédit à ce jour, qui se trouve à la Bibliothèque Nationale de France[2].

L'étude des manuscrits de l'édition et la très importante lettre à Juan de Vega nous indiquent que Maurolico a utilisé l'édition de Zamberti, sans négliger certains ajouts de Campanus[3]. Maurolico, comme à son habitude, réécrit les démonstrations, mais fait somme toute peu de changements significatifs. On aura une bonne idée de ces modifications avec la remarque que Maurolico écrit dans la marge supérieure du manuscrit au début du livre 9. La citation est complète : " Note que dans ce 9e [livre], les propositions 12, 16, 17, 24 et 25 sont ajoutées. Avec la 12e on démontre la 13e et la 14e plus facilement que dans les autres traditions. Avec la 16e et la 17e on démontre la 18e qui traite d'autant de [nombres] proportionnels que l'on veut, et non seulement de trois, comme dans les autres [traditions]. Et la 24e et la 25e traitent des moyennes proportionnelles, et cette considération n'était pas à négliger ".

L'ARITHMÉTIQUE NÉO-PYTHAGORICIENNE

L'arithmétique néo-pythagoricienne latine est largement dominée par les travaux de Boèce et de Jordanus. Maurolico a bien entendu travaillé ces textes et on sait qu'il en a fait un compendium et une édition. Ces travaux sont malheureusement perdus.

Le Sicilien abandonne les considérations philosophiques de Boèce et Nicomaque, et dans son introduction à son premier livre de ses *Arithmétiques*, fait le reproche à Jordanus d'avoir encore répété inutilement de nombreux résultats que l'on trouvait déjà dans les livres d'Euclide. Il se concentre uniquement sur les nombres figurés et revendique quelques résultats : de nouvelles démonstrations et la création de certains nombres figurés, en particulier ceux issus des solides réguliers.

Le changement par rapport à ces prédécesseurs latins est cependant plus profond. Le texte de Boèce ne comprend pas de démonstrations à proprement

---

2. F. Maurolico, *Euclidis elementorum [libri V,VII-X]*, manuscrit San Pantaleo 116/33 de la Biblioteca Nazionale V. Emanuele II de Rome, 118 folios ; *Compendium [Euclidis] in 10. elementorum libros*, manuscrit Fonds latin 7463 de la Bibliothèque Nationale de France, 59 folios.

3. F. Maurolico, lettre à J. de Vega du 8 août 1556, manuscrit Fonds latin 7473 de la Bibliothèque Nationale de France, folios 1r–16v ; éditée par F. Napoli, " Intorno alla vita ed ai lavori di F. Maurolico ", *Bullettino di bibliografia e di storia delle scienze matematiche e fisiche*, tome 9 (1876), 23–40.

parler, mais plutôt des explications et des constatations du résultat, souvent basées sur la construction et la représentation spatiale des nombres figurés. Jordanus, peut-être médiateur des travaux arabes sur le sujet, avait déjà éliminé les considérations philosophiques et fournissait des démonstrations numériques. Mais, dans les définitions comme dans plusieurs démonstrations, la formation spatiale jouait encore un rôle central. Le point de vue de Maurolico est encore plus radical. Le nombre figuré devient uniquement une entité numérique. Les figures ne servent plus que d'illustration. Les définitions sont purement numériques et les seules figures de ce livre — à une unique exception près, non significative pour ce qui nous intéresse — n'apparaissent que dans les marges des définitions. Les carrés sont les produits des racines par elles-mêmes, le n-ème nombre triangle est la somme des n premières racines, l'hexagone centré est la somme de l'unité et de six triangles de rang précédent, etc. Aucune propriété géométrique n'est mise en avant dans les démonstrations. On trouve par contre de nombreux tableaux de nombres figurés, établis jusqu'au rang 10.

Les extensions que promettaient Maurolico sont présentes. Jordanus avait introduit l'hexagone et l'octogone équiangles. Maurolico généralise l'idée à des figures qu'il nomme centrées ou du second genre, qui peuvent avoir autant de côtés que désirés. Mais il est surtout particulièrement fier de la création des nombres figurés issus des solides réguliers platoniciens : tétraèdres ou pyramides, octaèdres, hexaèdres ou cubes, icosaèdres et dodécaèdres. Il en donne une unique définition basée sur le nombre d'angles, d'arêtes et de faces, multipliées par des triangles et pyramides et montre que certains de ces solides pourtant définies de manières différentes, sont en fait numériquement égaux entre eux.

Dans ce contexte strictement numérique, où les différents objets que considère le mathématicien ont un rang, sont indexés sur les entiers naturels, où certaines définitions sont faites à partir de sommes ou de séries numériques finies, où des objets sont définis à partir d'autres objets, apparaissent des raisonnements mathématiques dans lesquels certains historiens ont voulu reconnaître la démonstration par récurrence. La situation est assez complexe et Maurolico utilise plusieurs méthodes de démonstrations différentes. Trois propositions, voire quatre, présentent rétrospectivement des ressemblances de forme avec notre méthode moderne de démonstration par récurrence. En particulier elles insistent sur une proposition-lemme permettant de passer d'un rang donné au rang supérieur, et répétée pour chaque rang. La proposition 15 est la plus souvent citée. Elle montre un résultat classique : les carrés s'obtiennent par l'addition des impairs successifs. Elle utilise pour cela une proposition clef — la 13e — qui permet de passer d'un carré, au carré du rang suivant par l'ajout d'un impair. Elle construit les carrés successifs jusqu'au rang 4 et se termine par la phrase : " Et ainsi de suite à l'infini, en répétant toujours la treizième [proposition], ce qui est proposé est démontré ".

Une lecture " inductive " voudrait qu'il n'y ait qu'une unique progression et que la démonstration soit globale, faite pour tous les entiers en une seule fois.

Mais il est indispensable d'examiner aussi quel type de démonstrations utilise Maurolico pour les autres propositions. La grande majorité est de type " quasi-général ", pour garder une terminologie introduite par Hans Freudhental[4]. Elles commencent très souvent par les mots *Exempli gratia*. Maurolico se place dés le début de la démonstration dans un cas particulier, le plus souvent le rang 5, ou 4. Puis il développe sa démonstration, en prenant toujours garde à ce que ce cas particulier soit facilement généralisable à un autre rang. C'est typiquement le cas des démonstrations euclidiennes. Le caractère de généralité est alors implicite : il est contenu dans le style de la démonstration, le cas particulier utilisé étant en quelque sorte générique. Il pallie, pourrait-on dire rétrospectivement, au manque de symbolisme et de calcul littéral. Pourtant Maurolico insiste sur cette généralisation et termine la moitié des démonstrations de la première partie de ce premier livre — exactement 33 sur 66 — par une phrase que l'on peut standardiser ainsi : ce qui a été fait pour le 5ᵉ — ou 4ᵉ — rang, peut être refait pour les autres rangs.

À la lumière de cette analyse du prototype de démonstration arithmétique chez Maurolico, il est possible de faire une lecture semblable des quatre propositions susceptibles d'accueillir la récurrence. Ce sont aussi des démonstrations quasi-générales. Elles sont démontrées dans un cas particulier — deux d'entre elles commencent elles aussi par *Exempli gratia*. Mais leur écriture est plus ambiguë. La phrase finale y joue le même rôle que dans toutes les autres propositions où on la trouve : expliquer qu'une démonstration semblable pourra être faite à un quelconque autre rang. Dans le cas de la proposition 15, la démonstration est faite uniquement pour le rang 4, cela de manière tout à fait générique. La dernière phrase qui faisait toute l'ambiguïté de la démonstration — " Et ainsi de suite à l'infini, en répétant toujours la treizième [proposition], ce qui est proposé est démontré " —, signifie que Maurolico rend explicite la nécessité, virtuelle bien sûr, de refaire la même démonstration pour tous les autres rangs. Elle ne signifie sans doute pas qu'il n'y a qu'une seule démonstration globale pour tous les rangs. C'est donc aussi à notre avis, une démonstration quasi-générale. Est-il encore nécessaire de remarquer que les dernières phrases des quatre propositions candidates à la récurrence chez Maurolico, sont les seules parmi toutes à contenir le mot " infini " ?

Le cas de Pascal est différent. En une occasion, le mathématicien français montre avoir reconnu dans le procédé de récurrence, une méthode de démonstration différente de celles habituellement utilisées : c'est la conséquence douzième — ou onzième selon l'édition — du *Traité du triangle arithmétique*[5]. En dehors de cette unique proposition, l'utilisation de l'induction chez Pascal est difficile à déceler, en tout cas est loin d'être systématique, et présente de gran-

4. H. Freudenthal, " Zur Geschichte Der Vollstandigen Induktion ", *Archives internationales d'histoire des sciences*, vol. 6, nᵒˢ 22–55, 17–37.

5. B. Pascal, *Œuvres complètes*, tome 2, texte établi, présenté et annoté par J. Mesnard, Bruges, Paris, Desclée de Brouwer, 1964–1992, 1187–1188 et 1294–1295, 4 vols.

des différences d'un cas à l'autre. Rappelons néanmoins ce qu'avait relevé G. Vacca (même si dans ce contexte, c'est maintenant à titre anecdotique) : Pascal cite nommément Maurolico à propos d'un résultat sur les nombres triangles dans une des lettres à Carcavi[6].

De toute façon, les arithmétiques de Maurolico semblent bien connues des mathématiciens français du XVIIᵉ siècle. Bachet de Méziriac en cite une proposition dans son appendice au *Livre des nombres polygonaux* de Diophante et cette proposition aura en outre l'honneur d'être commentée dans la marge de la fameuse édition que possédait Fermat[7]. Les nombres figurés continuent de jouer un rôle, sans doute plus marginal, mais non négligeable, dans les mathématiques françaises de l'époque. La palme revient à Frénicle de Bessy dans un travail jamais imprimé intitulé *Discours des nombres figurés*, aujourd'hui aux Archives de l'Académie des Sciences[8]. Frénicle trouve l'univers des nombres figurés trop complexe. Il cherche une nouvelle classification, critique les anciennes et opte pour une classification de type géométrique. Il s'intéresse en particulier aux règles de construction graphique des nombres figurés et sa classification repose sur ces règles. Mais il est amené à un tel nombre de catégories et d'exceptions qu'on ne peut pas dire que Frénicle réussisse son pari de réorganisation. C'est à propos de l'hexagone équiangle, qu'il nommera Maurolico, en en faisant l'inventeur. On trouvait déjà ce type de nombre figuré chez Jordanus.

L'ARITHMÉTIQUE DES QUANTITÉS GÉNÉRALES

Le moyen âge et la Renaissance latins sont riches de travaux arithmétisants des irrationnels du livre 10 des *Éléments* d'Euclide. Fibonacci, Pacioli, Cardan, les cossistes allemands, mais aussi Campanus avec son édition d'Euclide, y ont tous contribué. Dans la plupart des livres d'arithmétique de la Renaissance, les liens entre nombres irrationnels et lignes euclidiennes du livre 10 et le statut même de ces nombres dans la mathématique sont rarement explicités. Si l'on considère le panorama des quantités irrationnelles dans la Renaissance, on a d'un côté le toujours solide livre 10 des *Éléments* d'Euclide, et de l'autre un foisonnement de travaux arithmétiques qui manipulent des nombres dont certains, assez ésotériques, peuvent être " associés " sans vraiment de justification aux lignes euclidiennes. On peut considérer en première approximation le deuxième livre des Arithmétiques de Maurolico comme participant à une

6. G. Vacca, " Sur le principe d'induction mathématique ", *Revue de métaphysique et de morale*, t. 9 (1911), 30–33.

7. Bachet de Méziriac, *Diophanti alexandrini de multibus numeris. Appendix ad librum de numeris polygonis*, Paris, 1621, 55 ; P. de Fermat, *Œuvres de Fermat*, tome I, publiées par les soins de MM. Paul Tannery et C. Henri, Paris, 1891, 341 et 342.

8. Frénicle de Bessy, *Discours des nombres figurés*, manuscrit des Archives de l'Académie des Sciences, fond Roberval, carton 4, document 14. Je dois à C. Goldstein l'information de l'existence de ce manuscrit.

lignée d'arithmétiques pratiques qui traitent des irrationnels euclidiens sous forme de nombres. Maurolico veut en quelque sorte réunir deux mondes qui s'ignorent et donner un cadre théorique à ces arithmétiques pratiques qui traitent d'irrationnels euclidiens. L'idée de Maurolico paraît déjà assez ambitieuse.

Toutefois l'objet qu'il traite n'est pas le nombre mais ce qu'il nomme " la quantité générale ", sorte de lieu commun entre la quantité discrète et la quantité continue dont le vecteur, " l'instrument ", comme il le dira, est le nombre. L'enjeu épistémologique est différent. D'un ouvrage d'arithmétique de la Renaissance, on passe à un travail particulièrement original lié à la *Mathesis universalis*. Proclus avait lancé l'idée d'une mathématique commune et universelle parce que basée sur des principes communs à tous les genres de mathématiques. Avec les termes exacts de *Mathesis universalis* et dans sa quatrième des *Règles pour la direction de l'esprit*, Descartes est parvenu à un certain aboutissement de ce concept. Giovanni Crapulli en a cherché les sources à la Renaissance en insistant particulièrement sur le Belge Adrian Van Roomen[9]. Maurolico a lui aussi dignement sa place dans l'histoire de la *Mathesis universalis*.

Le 2e livre des arithmétiques de Maurolico semble s'intéresser moins au nombre irrationnel en tant que représentant de la ligne continue des irrationnels euclidiens du livre 10, qu'au nombre en tant que vecteur d'une théorie des " grandeurs générales ". Pour le Sicilien — et ce que nous disons maintenant ressort des prolégomènes de ce deuxième livre d'Arithmétique, du texte même bien sûr, mais aussi d'un passage du *Sermo de quantitate*[10] —, la quantité générale est l'objet de la mathématique première — *primaria mathematica*. Les résultats communs aux quantités discrètes et continues sont avant tout des résultats de la quantité générale. Les opérations simples comme addition, soustraction, multiplication, division, mais aussi les manipulations de rapports, l'irrationalité et la commensurabilité, peuvent être le fait aussi bien de la quantité continue que de la quantité discrète. Elles sont en fait l'apanage de la quantité générale. Celle-ci est en quelque sorte, leur intersection, leur quintessence, et en ce sens elle leur est supérieure, " plus digne, plus pure ", comme le dira Maurolico. Surtout, son instrument — *instrumentum* —, son vecteur, est le nombre. Son mode opératoire, les démonstrations qui la mettent en jeu, son vocabulaire, sont de types arithmétiques.

Pour pouvoir caractériser la quantité générale ou plus exactement savoir si un résultat est le fait d'une quantité particulière, discrète ou continue, ou bien le fait de la quantité générale, il est donc nécessaire que ce résultat soit " commun " aux deux quantités particulières. Maurolico ne peut pas montrer

9. G. Crapulli, *Mathesis universalis, genesi di una idea nel XVI secolo*, Roma, Edizioni dell'Ateneo, 1969.

10. F. Maurolico, *Prologi sive sermones. De divisione artium. De quantitate. De proportione*, in G. Bellifemine (ed.), Melphicti, 29 et 30.

cette affinité, cette équivalence, mais il peut la faire constater au cas par cas. C'est ce qu'il fait en particulier dans la première proposition de ce deuxième livre : " Tout ce que nous avons montré sur la multiplication, le rapport, la proportion, la symétrie et la similitude des nombres, lignes et solides, nous pouvons de même le conclure et le démontrer pour n'importe quel genre de quantité ".

Cette proposition est une des affirmations centrales de la thèse de Maurolico sur la réalité de la quantité générale. La proposition n'a d'ailleurs pas uniquement ce rôle de justification : elle a aussi un rôle infiniment plus prosaïque. Elle signifie aussi que dans le cours des démonstrations de ce deuxième livre, Maurolico pourra faire appel aux résultats de nombreuses propositions euclidiennes — celles qui seront communes aux quantités discrètes et continues — parce qu'elles s'appliqueront aussi à la quantité générale, pour ses propres résultats.

La démonstration de cette proposition n'en est pas vraiment une ! Maurolico donne cinq exemples de " traduction " ou de " transfert " — ce sont encore les termes qu'il utilisera — de propositions et de leur démonstration, de la quantité particulière, discrète ou continue, à la quantité générale. Les exemples concernent les propositions 17 et 18 du livre 7 des *Éléments* d'Euclide — dans la version de Campanus —, la première proposition du livre 2 et deux propositions du livre 10. Après chaque exemple particulier, on trouve une généralisation du type : tout ce qu'Euclide a démontré dans le livre 2 [ou 5, ou 6, ou 7, ou 10, donc dans le cas d'une quantité particulière], peut être démontré pour la quantité générale. Quand Maurolico pense avoir donné suffisamment la preuve de sa proposition, il conclut : " Donc par cette proposition, toute spéculation géométrique est ramenée à la pratique numérique " — la spéculation arithmétique reste ce qu'elle est. La suite de la première partie a pour objet les opérations de base sur les quantités générales : addition, soustraction, multiplication, division, racines carrées, cubiques. La quantité générale est toujours " signifiée " par un nombre. Les démonstrations sont faites dans chacun des cas où la quantité générale est signifiée, par un entier, une fraction, un irrationnel simple, du type racine carrée, cubique, etc. et un irrationnel composé à la façon des irrationnels du 10e livre des *Éléments*. Maurolico montre comment effectuer les opérations de base selon le ou les nombres qui signifient la quantité. La deuxième partie est spécialement consacrée aux irrationnels euclidiens du livre 10 dont on retrouve bon nombre de résultats. On retombe ainsi sur un autre aspect du travail de Maurolico : l'arithmétique pratique et la " démonstration de la pratique à partir de la théorie ". Maurolico nous dit bien vouloir " démontrer la pratique numérique [des irrationnels] ", mais il n'est pas plus explicite sur ce qu'il entend dans ce cas précis. En tout cas, le formalisme que Maurolico crée et les propositions qu'il démontre dans son livre, permettent dans une large mesure de justifier les nombreuses règles et les exemples que l'on trouve dans les livres d'arithmétique pratique qui traitent d'irrationnels à la Renaissance.

C'est dans le cadre des mathématiques et en particulier de l'arithmétique du moyen âge et de son époque que Maurolico s'avance sur le terrain de ce qu'il appelle mathématique première. L'horizon de Van Roomen est différent. Dans *l'Apologia pro Archimede* de 1597[11], il s'agit avant tout de défendre Archimède contre les attaques de Scaliger, qui accusait scandaleusement l'illustre Syracusain d'avoir utilisé dans *La mesure du cercle* des nombres pour résoudre un problème géométrique. Van Roomen contre-attaque et cherche à montrer que l'arithmétique peut être utilisée en géométrie et vice-versa. Et quelle preuve n'aurait-il pas, s'il crée une mathématique qui dépasse les particularités de la géométrie et de l'arithmétique, dans laquelle il n'est plus possible de les différencier et où la question de l'utilisation de l'une dans l'autre n'a plus de sens. Van Roomen utilise pour ce faire une multitude d'arguments. Nous en retiendrons deux. En premier les arguments d'autorité. Van Roomen cite une douzaine de mathématiciens de la Renaissance qui ont mêlé nombres et grandeurs géométriques dans les démonstrations. Le dernier dans la liste est Maurolico : " Et incontestablement encore de nos jours, Maurolico, abbé de Messine et mathématicien renommé, a rédigé explicitement son deuxième livre pour qu'il montre de quelle manière on trouve les dimensions des grandeurs. Ce livre est assurément très utile et peut dans son entier confirmer notre manière de voir. Il suffit en vérité de citer une partie de la préface, pour qu'apparaisse clairement le propos de tout le livre. Cet homme très savant écrivait … " .

Van Roomen cite alors textuellement la première moitié du premier prolégomène du deuxième livre des arithmétiques du Sicilien — soit une quinzaine de lignes. Il n'est peut-être pas indifférent que Maurolico soit le seul mathématicien contemporain à obtenir la faveur d'une si longue citation.

Le deuxième élément décisif pour la pensée du Belge est une proposition d'Eutocius dans son commentaire aux *Coniques* d'Apollonius, commentaire à la proposition 11 du premier livre portant sur la composition de proportions. La démonstration d'Eutocius est purement arithmétique. Eutocius justifie ce caractère arithmétique par des arguments très différents : les Anciens utilisaient eux aussi des nombres pour des démonstrations sur les proportions ; ces démonstrations sont plus mathématiques — dans un sens très général — qu'arithmétiques ; les proportions résident en premier lieu dans les nombres, puis dans les grandeurs ; enfin, les disciplines mathématiques sont cousines. Van Roomen commente un à un tous ces arguments. Il en tire l'existence " d'une science mathématique commune à l'arithmétique et la géométrie, vers

---

11. A. Van Roomen, *In Archimedis circuli dimensionem expositio et analysis. Apologia pro Archimede ad clariss. virum Iosephum Scaligerum. Exercitationes cyclicae contra Iosephum Scaligerum, Orontium Finaeum et Raymarum Ursum in decem dialogos distinctae*, Wurceburgi, 1597.

laquelle regardent les propriétés communes à toutes les grandeurs… " et il conclut sur la nécessité des nombres pour cette science commune. Dans les chapitres suivants, Van Roomen cherchera à créer une ébauche de *Mathesis universalis*, avec ses définitions, ses axiomes et quelques propositions. L'arithmétique y joue un rôle déterminant, puisque c'est par elle que se font les démonstrations, dans lesquels les grandeurs et les proportions sont " expliquées " par des nombres. Le Belge donnera encore quelques années plus tard une classification de la mathématique[12]. La mathématique pure se divise entre universelle et particulière. La particulière est constituée de façon banale, de l'arithmétique et de la géométrie. La mathématique universelle se divise en mathématique première et logistique — ou *supputatrix*, ou *arithmopraxis*. La logistique " tire ce qui est cherché des nombres accommodés aux choses ", et est l'instrument de la mathématique première : " Et ce que la logique est à la philosophie universelle, on peut estimer que la logistique l'est à la mathesis ".

On remarquera maintenant les nombreux points communs que présentent les idées de Van Roomen et de son prédécesseur Maurolico. Tous les deux présentent une " Mathématique première ", dont l'objet est la quantité " absolue ", dira le Belge, " générale " dira Maurolico, dont les théorèmes sont communs aux sciences particulières, et surtout dont l'instrument est un certain type d'arithmétique, " logistique " pour l'un, " arithmétique pratique " pour l'autre. Une telle communion d'idées laisse penser que les deux mathématiciens se seraient encore accordés sur le vocabulaire déjà très proche. On notera les origines a priori diverses des deux constructions. Maurolico trouve son bonheur dans les mathématiques médiévales et contemporaines, à la fois dans les nombreux liens qui se sont tissés entre nombres et grandeurs, et dans la pratique arithmétique des irrationnels euclidiens. Van Roomen centre son discours sur une proposition d'Eutocius dont il tire toute la moelle, et d'arguments d'autorités, d'Euclide aux mathématiciens et arithméticiens contemporains, parmi lesquels on ne peut que remarquer Maurolico.

---

12. A. Van Roomen, *Universae mathesis idea, qua mathematicae universim sumptae natura, praestentia, usus et distributio brevissime proponuntur*, Herbipoli, 1602.

SOME ASPECTS OF MAUROLICO'S OPTICS

Romano GATTO

Maurolico's works of optics were published posthumously, for the first time, in Naples in 1611 collected in a single volume.

The book consists of 83 pages in all, most of them devoted to the *Photismi* and to the *Diaphanarum* and I will speak only about these. The Neapolitan edition was edited by Christoph Clavius and his scholar Giovanni Giacomo Staserio. Clavius inserted in the text some personal explanations and short comments which, he thought, had completed and clarified some aspects of Maurolico's exposition.

In 1613, two years later, a second edition of Maurolico's book was printed. It is the revised and checked edition of the previous one, probably written making use of the same text. The unique substantial difference in this edition is the presence of some marginal bibliographic notes not included in the previous one, and not ascribed to Maurolico. We are of the opinion that these notes were added by the editor, uniquely to highlight the authors who had dealt with the same matters that had been treated by Maurolico, rather than to indicate Maurolico's sources. Really the question of Maurolico's sources is more complex, and it constitutes a chapter in my studies which I haven't closed.

The first relevant aspect of the *Photismi* is the approach essentially geometric by which the author deals with all the matters. Indeed the medieval Optics, which was still very relevant in the 16th century, was an essentially empirical discipline in which the experimentum often had a demonstrative role. The use of geometry, present in it too, had pre-eminently an instrumental character because it was essentially directed towards the description and representation of the phenomena. It is true that there were some treatises as Περὶ ὀπτικῆς by Witelo[1], whose form was structured according to the model of Euclid's *Elements*, with an apparatus of axioms and definitions and with some series of propositions and theorems ; but it is also true that this formal order wasn't followed by a corresponding systematic use of the deductive Euclidean method[2].

On the contrary, Maurolico's Optics, being a discipline based on the observation of empirical phenomena, uses Euclide's *Elements*, not only as a formal model, but also as a theoretic support and an instrument to give his theories character of deductive science. In Maurolico's Optics the terms — definition, supposition, theorem and corollary — have the meaning of elements constituting a geometrical theory.

Maurolico doesn't conceive the demonstration out of the geometrical order. Only in a few cases he swerves from his geometric line, including among the theorems some enunciation of questions he would insert better among the definitions or the postulates since nothing is demonstrated in them at all, but are only given enunciations of matters which are justified by the empirical experience. We must regard these cases, which are very rare, as a limit of a statement, that, in order to be completely geometrical, ought to suppose the axiomatization of all the physical properties that occurred in the treatment.

At the beginning of the *Photismi* Maurolico puts forward only the following 4 definitions :

1. *Lucidorum aliud quidem per se radiat, ut Sol, flamma ; aliud autem aliunde receptum lumen, reflectit, ut Luna, speculum*

2. *Primariam ergo lucem vocabimus eam, quae immediate a corpore per se radiante procedit*

3. *Eam vero quae ex prima, vel quotacumque reflexione fit, secundariam dicemus*

4. *Umbram quoque appellabimus vel universalem, vel particularem luminis absentiam*

For Maurolico *lucidus* is, not only a body which irradiates its own light, but is also a body which irradiates reflected light. According to this fact, in the theory of the reflection of light, the points in which the reflection takes place gain the same properties of the source points. So the difference between the primary light and the secondary light is only a nominal difference. Moreover the 4th definition announces one of the most important results obtained by Maurolico about the penumbra, a phenomenon not studied until that time which I will speak about later.

Six axioms (*Suppositi*) follow the four definitions :

1. Witelo, Περὶ ὀπτικῆς *de natura, ratione et proiectione radiorum visus, luminum colorum atque formarum quam vulgo perspectiva vocant, libri X... Nunc primum opera Mathematicorum praestantis d. Georgii Tanstetter et Petri Apiani in lucem aedita*, Nürnberg, 1535.

2. Together with some theorems geometrically demonstrated some of them are only justified experimentaliter ; others which derive from considerations that have a little regard to Optics ; at last some others in which nothing is demonstrated, but they are expounded simply explained some theories.

1 - *Omne lucidum punctum per rectam radiare lineam*

2 - *Densiores radios intensius : aeque vero densos aequaliter illuminare*

3 - *Ab uno speculi puncto, in quod signum lucidi quodpiam irradiat, in unum quoque reflexionem fieri*

4 - *Lucido ad illuminati, illuminato usque ad lucidi locum traslatis, lucidum adhuc eodem, quo primus, tramite ad illuminatum radiat*

5 - *Plures radios intensius : aequales vero aequaliter illuminare*

6 - *Videtur deesse hoc principium : Ab angulis aequalibus aequales numero radij emittuntur, et a maiori plures : propterea quod radij debent esse inter se distantes determinato quodam modo. Non enim sub quocumque angulo res videtur*

Indeed Maurolico established only the first five axioms ; the 6th, was added by Clavius.

According to current ideas, the first axiom sanctions the rectilinear propagation of light. The 2nd and the 5th axioms introduce the concept of density without defining it, but only characterising it by means of some properties. But by the context of the treatment it is no doubt that for Maurolico density of rays is the number of rays for unity of surface (or, in a plane representation, for unity of segment). The 3rd axiom establishes that a luminous point is reflected only in one point. The following 4th axiom says that, if the place of the luminous point source is exchanged with the illumined one, the radiation takes place along the same way as previously. These two axioms, which establish the possibility to exchange the incident ray with the reflected one, together, constitute what I call a *reciprocity principle* which Maurolico utilises several times as demonstrative technique. The 2nd and 5th axioms together constitute what I call a *comparison principle* (a metric of a geometrical system), which Maurolico will utilise in developing his geometrical theory of more or less illumination's phenomena, which represents the first real approach to a systematic study of photometry, a field no more explored at that time. In fact, by means of these two axioms, Maurolico can establish a correspondence between the luminous rays, and the elements of surface (or of line) on which the same rays fall. And as it is a correspondence between physical entia and geometric entia, he reduces the resolution of the photometric process of more or less illumination to the comparison of surfaces or of segments. Then Maurolico deals with the theory of the shade of bodies illuminated by a point source considering the projection of plane figures and of the spheres. At the end of this theory he passes to the consideration of bodies which are illuminated by a wide source, and he completely characterises the concept of shade given in the 4th Definition, not only as a total lack, but also as a partial lack of light. So he distinguishes the case of proper shade, in which no ray coming from the luminous source falls on a region of the plane (total lack), by the penumbra's one in which only some rays coming from the luminous source fall on a region of the plane (partial lack). This last phenomenon, which was of a difficult interpretation, was until that moment nearly unknown by Optics' authors. Maurolico understood it directly depended on the rectilinear propagation of the light which didn't permit all the rays of the wide source to illuminate in the same way the regions nearest to the shade zone.

But let us follow Maurolico in the explanation of the XVIIIth theorem : *Quo maius fuerit lucidum, quoque magis illuminatum a plano, in quod umbra proijcitur, destiterit, eo maiores atque intentiores umbrae termini videntur.*

FIGURE 1

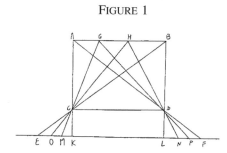

Let AB a wide luminous body and CD an opaque body which projects a shade on the surface EF. Maurolico shows that the plane regions beyond the points E and F are completely illuminated, the one included between K and L is completely in the dark, while the regions AK and LF, in which there is a progressive passage by the total shade to the total light, are penumbra's zones. In fact these zones are illuminated by the rays coming out only from a part of the illuminated body. Indeed MK is only illuminated by the rays coming out from AG and OM from the ones coming out only from AG and from GH. And so on till arriving at E, at a point beyond them it will be completely light. Therefore, passing from K to E the illumination increases because the number of the rays which arrive on the surface continuously increases. Similarly for the zone LF.

One of the most interesting subjects of the *Photismi* is the treatment of the reflection. Most of medieval authors dealt with this phenomenon essentially by experimental way (only in Bacon's *Opus Maius* we find a partial geometric consideration of this phenomenon, and what's more, by means the supposition a priori of some hypotheses which removes the generality of the question). On the contrary Maurolico deals with the reflection with the greatest generality and rigorously by geometry, as it was really a geometric problem, making use of the above mentioned reciprocity principle.

First of all he demonstrates that (Theorema XXVI) : *Perpendicularis in speculum radius in seipsum reflectitur.*

FIGURE 2

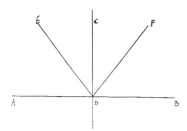

In fact, if ab absurdo it was reflected in the way of DF making the angle CDF, called CDE the symmetric angle of CDF with respect to the vertical CD, it would result CDF=CDE. Then the ray CD would be reflected on E and on F against the third *suppositum*. Therefore the reflection must only follow in the same way as DC.

Soon after Maurolico considers the case of an incident ray oblique with respect to the reflecting plane. He shows that, in this case (Theorema XXVII) : *Obliquus in speculum radius ad aequalem inclinationis angulum in plano ad speculum recto reflectitur.*

FIGURE3

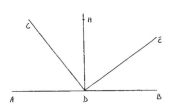

Also this demonstration is made ab absurdo. Firstly Maurolico supposes that the rays of incidence and of reflection belong to the same plane perpendicular to the reflecting one, but that the reflection angle BDE isn't equal, but smaller than the incidence angle ADC, that is to say that the reflection of the point C happens on the point E with an angle smaller than the one of incidence. Then exchanging the role of the luminous point C with the reflected point E, by the IVth *suppositum* (*reciprocity principle*) the reflection will happen in the way EDC at the point C and, by the made hypothesis, the angle BDE will be wider than the angle ADC. But it is just the contrary of that previously supposed ; that is absurd. Therefore ADC cannot be wider than BDE. And, as analogously it is possible to demonstrate that ADC cannot be smaller than BDE, it is necessary that ADC=BDE.

At this point Maurolico shows that if, contrary to what he supposed, the reflection ray doesn't belong to the plane of AB and CD, that is to say that the reflected point of C is F not belonging to the plane of AB and CD, and such that the ray DF is equally inclined as DE in respect to the plane AB, as FDC must be equal to EDC, there it would be reflection on E like on F, in opposition to the 3rd *suppositum* which states the uniqueness of the reflection's point[3].

Maurolico gives a complete treatment of the reflection besides considering the reflection of plane mirrors but also of convex and concave mirrors.

Diaphaneon's text is divided into three books ; the first is devoted to the refraction's theory, the second to the iris' theory ; the third to the physiologic theory of eye and that regarding the processes of vision. As in the case of the *Photismi*, the author's intention is to deal with all these subjects geometrically, but sometimes, because of the same nature of the matter he considers, he is obliged to resort to results deduced by experience.

The first book is introduced by a little theoretic apparatus : one *Definitio* and four *Supposita*. Really, the definition is rather anomalous. In fact, nothing is defined in it. It is merely the description of an experience exposed by Euclid and consisting in the observation of an object, which is inside a transparent vessel, having the eyes perpendicular to it, in two different moments : when inside the vessel there is water, and when there is not water.

The four *Suppositi* derive just from this experience.

1. *Perpendicularem radium in diaphanum recte procedere ; obliquum vero versus perpendicularem frangi*

2. *Radios aeque inclinatos, aeque frangi ; magis vero inclinatum, magis*

3. *Multiplicato angulo inclinationis, angulum quoque fractionis aequaliter multiplicari*

4. *Rem apparere in loco concursus radij visualis recti cum ea, quae ab re ipsa in planum diaphani, perpendicularis progreditur*

FIGURE 4

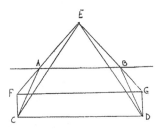

3. By the same way, Maurolico shows that the reflection ray and the incident ray belong to the same plane perpendicular to the reflecting one.

The first supposition expresses a necessary condition in order to verify the refraction : a ray perpendicular to the surface of a transparency always proceeds according to a straight line without any deviation ; on the contrary, an oblique ray, passing through the transparency, deviates in the direction of the perpendicular. The second and the third suppositions give a quantitative characterisation of the phenomenon. In particular, the second one establishes that equal incident angles correspond to equal refraction angles, while wider incident angles correspond to wider refraction angles. The third one says that incident angles are in proportion with the corresponding refraction angles. At last, the fourth supposition expresses the author's opinion about the image's formation by the refraction. If we consider an object FG below the transparency AB and the rectilinear visual rays EF and EG, by refraction these rays are refracted in the way of EAC and EBD respectively. Then if we draw the perpendicular lines to FG by the extremes F and G, these lines intersect the refracted rays EAC and EBG in the points C and D which represent the extremes of the image of FG.

Starting from these principles, in the first nine theorems, Maurolico gives a wide description of the refraction by transparent means of different form. I cannot dwell upon them, because I haven't time. But I want to call your attention to the Xth Theorem which enunciates the following : *Anguli inclinationum sunt fractionum angulis proportionales.*

In fact, it could seem that, by this theorem, Maurolico wanted to demonstrate the assertion previously given as an axiom, in the 3rd supposition. Indeed, reading the demonstration of this theorem, we realise it would be enunciated differently. To understand that it is necessary to remark that Maurolico in his suppositions speaks of inclined rays, inclination angles and refraction angles without previously giving any definition of them.

FIGURE 5

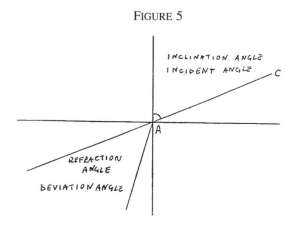

But, by the context of his treatment, and particularly by the demonstration of the IXth Theorem, it is not difficult to understand that what he calls inclination angle is the incident angle, that is the angle formed by the incident ray the transparency's surface and the perpendicular to the same surface in the incident point, while refraction angle is the one formed by the refracted ray and by the prolongation of the incident ray that is quite different from the current acceptation according to which refraction angle is the angle formed by the refracted ray and the perpendicular to the transparency's surface in the refraction's point. Today, Maurolico's refracted angle could be defined as an angle of deviation from the direction of the incident ray. Then, according to this last definition, the 3rd supposition assumes the following enunciation : " making a multiple of the inclination angle, the deviation angle becomes multiple in the same way ", that it is to say, " the deviation angles are proportional to the inclination ones ". So defined the deviation angle is the difference between the inclination angle and the refracted angle ; therefore it is strictly joined to the refracted angle[4]. Indeed, in all of his remaining treatment, Maurolico continues to consider refraction angle as the one we called deviation angle. Particularly, the misunderstanding of the enunciation of the Xth Theorem is completely dissolved by the demonstration of the Corollary to the XIXth Theorem. In fact, there the author calls the " angle contained between the perpendicular and the refracted ray " the angle which, according to the current acceptation, is the refraction angle, differentiating it from the one formed by the refracted ray and the prolongation of the incident ray that he continues to call refraction angle. Therefore, if we want to preserve the meaning given by Maurolico to the refraction angle, we must enunciate the Xth Theorem as follows : " The inclination angles are proportional to the ones formed by the refracted ray and the perpendicular ", that is quite a different thing from the 3rd supposition.

Then Maurolico considers the refraction in a transparent sphere to introduce the theory of the telescope. In fact, he utilises the transparent sphere as a system of two joined lenses, the one concave and the other convex. So he explains the various phenomena which constitute the principles of such an instrument.

As we previously said, the second book is devoted to the iris' theory. It is important to remark that, as Maurolico affirms, he was the first who dealt with this theory completely by geometry. By this way he explains the nature, the form, the greatness and the colours of the iris.

In the XXVth Theorem Maurolico establishes the conditions which determine the iris' formation. He demonstrates that iris was generated by the refraction

---

4. So, if we give as an axiom the enunciation of the Xth Theorem (inclination angles are proportional to the refraction's ones), we could demonstrate that Maurolico supposed as axiom in the 3rd supposition, that is : deviation angles are proportional to the inclination's ones.

to 45° undergone by the suns' rays passing through the dew drops. In an *Additio* to this theorem, Maurolico specifies that this phenomenon is not only due to the refraction, but also to the reflection. In fact the refraction by itself wouldn't be able to generate and make the iris visible. Indeed, because of the tiny dimensions of a drop, the luminous ray would be too faint and couldn't arrive to the eye. To make visible the iris the refraction is joined to the continuous reflection of the sun ray on eight points of the internal surface of the drop. These points constitute the vertexes of an octagon whose diagonals are the reflected rays. Only after such iterated reflections inside the drop, the ray comes from it carrying a lot of light, which added to the one carried by the rays coming from the other drops, permits the formation and the vision of the iris. In some following theorems and *Additions* Maurolico completely characterises the iris and the conditions under which this phenomenon is visible.

FIGURE 6

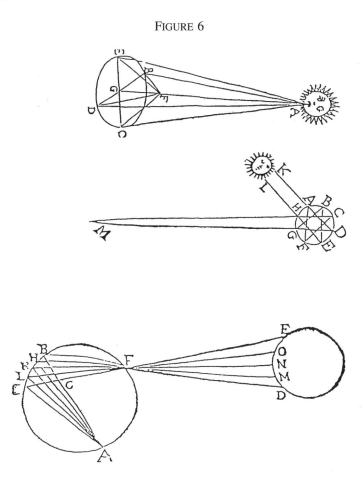

As regard to the colours Maurolico demonstrates that the difference of the colours is due to the different density of rays in the bands constituting the iris, and that the reason of this is the bending of the iris. He also explains why sometimes it is possible to see a secondary iris together with the primary one and why the colours of this secondary iris are ordered in the contrary way with respect to the ones of the primary. He confutes a largely diffused opinion at that time, according to that the secondary iris was the reflected image of the primary one. He demonstrates that the generation of the secondary iris happens in the same way as the primary one, and that it is independent of the first iris.

I want to conclude that the demonstration of these theorems represents the first real successful attempt at scientific description of the formation of this phenomenon, which is still appreciable.

LA DOCTRINE DES SECTIONS DU CÔNE APPLIQUÉE À LA GNOMONIQUE CHEZ F. MAUROLICO

A.C. GARIBALDI

1. Etienne Montucla, dans son *Histoire des Mathématiques*, a donné un abrégé d'histoire de la gnomonique[1]. Après avoir parlé de la gnomonique des anciens, il passe à celle des modernes en affirmant : " aux yeux du géomètre intelligent elle ne consiste qu'en quelques problèmes peu difficiles ".

Il expose tout de suite un schéma géométrique très simplifié pour résoudre ces problèmes.

Après avoir rappelé quelques traités arabes, Montucla dit : " la gnomonique renaquit aussi en Europe avec l'astronomie ". Il cite Werner et d'autres auteurs du XVe siècle qui sont restés manuscrits. En passant au XVIe siècle, il rappelle Münster, Oronce Finé (corrigé par Nonius), G.B. Vicomercato, Commandin[2]. Il donne ensuite un long catalogue d'oeuvres de gnomonique des XVIIe et XVIIIe siècles.

A ce moment là Montucla présente et souligne une distinction fondamentale : " quelques géomètres ont envisagé la gnomonique d'une manière plus savante, et n'ont pas dédaigné d'y appliquer l'analyse et des considérations même de la géométrie transcendante. Maurolicus en avoit donné l'exemple dans son traité *De lineis horariis* en remarquant que les arcs des signes des mois solsticiaux sont des sections coniques, que les heures babyloniennes et italiques sont des tangentes à une section conique déterminée "[3].

---

1. Dans le supplément au quatrième livre de E. Montucla, *Histoire des Mathématiques,* tome I, 1799, 725-735.

2. Nous donnerons ici seulement les références aux oeuvres nommées par Maurolic, c'est-à-dire Münster et Commandin. S. Münster publia en 1531 le livre : *De omnium generum horologiorum compositione et fabricatione.* Ce texte parut aussi *tertiò recognitus et locupletatus* en 1551 dans les *Rudimenta mathematica* du même auteur. On ne sait pas à quelle édition se référait Maurolic. F. Commandin, à la suite de son édition de l'Analemme de Ptolémée (1566), publia le texte *de horologiorum descriptione*, que Maurolico critique dans son texte de 1569.

3. E. Montucla, *Histoire des Mathématiques, op. cit.,* 734. Les " arcs des signes " y sont expliqués, 727.

Nous reviendrons plus loin sur ce qui est la nouveauté de Maurolico. Maintenant, rappelons avant tout que l'oeuvre *De lineis horariis libelli tres,* la plus complète que Maurolico dédia à ce sujet, fut publiée en 1575 dans les *Opuscula Mathematica,* comme d'habitude, avec le colophon qui indiquait la date du manuscrit passé à l'imprimeur : 1553. Les *Opuscula* contiennent aussi un petit abrégé, avec un but essentiellement pratique, *De lineis horariis brevis tractactio* qui est daté 1569[4]. Le fondement théorique de l'ensemble avait été donné bien auparavant dans la *Cosmographia* parue en 1543[5]. On trouve aussi énoncées les propositions fondamentales de la Gnomonique dans le *Compendium Mathematicae* publié en 1558 à la suite des *Sphériques* de Théodose éditées par le même Maurolico[6].

Pour ce qui regarde la gnomonique de son temps, Maurolico écrivait dans la *Brevis tractatio* de 1569 : *(…) loquamur de lineis horariis. Haec enim est compendii nostri materia. De his recentiores quidam scripsere. Sebastianus quidam (Münster) fabricam earum tradidit : sed speculationem neglexit. Federicus noster Urbinas (Commandin), dum theoriam nimis affectat, obscure locutus est… Nos autem rem ipsam tribus olim libellis complexi sumus, fundamentum theoriae et praxim exponentes (…)*[7].

Il se réfère donc à son oeuvre majeure sur la gnomonique en trois livres. Dans le second de ces livres, en s'adressant à Vega, il avait défendu son choix pour la théorie en disant : *Opere precium igitur facturus videor, et rem speculativis ingenijs gratissimam, si huiusmodi periferiarum proprietates et formas, quantum ad ipsum spectat negocium, hic exequar : quod cum ex conicorum elementorum doctrina pendeat, et ad subiecti theoriam magis, quam ad praxim pertineat ; ab his, qui de horologijs huiusmodi scripserunt, quos ego sciam, hactenus neglectum est*[8].

La théorie qui est nécessaire pour la gnomonique dépend donc de la doctrine des coniques. Au moment où Maurolico écrit son traité, les premiers quatre livres d'Apollonius sont disponibles en latin dans la version de Memo[9], un texte très mauvais que Maurolico avait amélioré en 1547, sans pourtant réussir à le publier[10]. On comprend alors la raison pour laquelle il donne dans son traité un résumé — esquisse sur la doctrine des coniques, limité à ce que lui

---

4. Ces deux textes se trouvent dans les *Opuscula Mathematica,* Venetiis, 1575 respectivement 161-285 et 80-102.

5. La *Cosmographia,* dédiée à Bembo, fut publiée à Venise, 1543. On y parle de la gnomonique à cartes 60-61.

6. Cette oeuvre : " Theodosii Sphaericorum *Elementorum* libri III ex traditione Maurolyci… " parût à Messine, apud Petrum Spiram, 1558.

7. *Opuscula Mathematica, op. cit.,* 80.

8. *Opuscula Mathematica, op. cit.,* 211.

9. Cette traduction des premiers quatre livres des *Coniques* parût à Venise, en 1537.

10. L'exposition de Maurolico paraîtra seulement en 1654 : elle contient les premiers quatre livres d'Apollonius et une restitution (*divinatio*) des livres V-VI qui n'étaient pas encore disponibles.

est nécessaire. Ce résumé, qui forme le troisième livre du *De lineis horariis,* est l'objet la présente communication.

2. Avant tout, cependant, il faut rappeler quelques notions de gnomonique en suivant Maurolico. On considère d'abord la sphère céleste qui tourne chaque jour autour de l'axe du Monde[11]. Sur cette sphère on a une première espèce de cercles horaires qui passent par les pôles de l'équateur et commencent au Midi (c'est-à-dire du cercle méridien) : ces cercles nous donnent les 24 heures qu'on appelle astronomiques. On obtient une autre espèce de cercles horaires en envisageant les deux parallèles qui séparent, en un lieu donné avec son horizon, les choses visibles de celles invisibles. On peut alors mener sur la sphère les cercles qui touchent ces parallèles en partant du coucher ou du lever du Soleil : ils sont aussi en nombre de 24, équidistants l'un de l'autre et donnent les heures qu'on appelle babyloniennes ou italiennes[12].

Le plan de l'horloge solaire en coupant les cercles horaires donne lieu à deux espèces de lignes horaires du même nom. Mais, comme les deux parallèles qui séparent les choses visibles de celles invisibles forment les bases de deux cônes qui ont le même sommet au centre du Monde, le plan de l'horloge coupera ces cônes dans un cercle ou dans une des sections coniques. Or, les lignes horaires de la première espèces, qui partent du Midi, coupent ces coniques en des points où les lignes horaires de l'autre espèce, qui partent du coucher ou du lever du soleil y sont tangentes. On peut donc, en suivant Maurolico, nommer " sécantes " les lignes horaires de la première espèce et " tangentes " celle de la deuxième.

Tandis que les praticiens construisaient les lignes horaires dans le plan de l'horloge par points en partant de l'observation, c'est-à-dire en observant directement le style, ou gnomon, et l'ombre qu'il projette sur le plan de l'horloge, Maurolico profite du lien avec les coniques, pour donner la construction.

Le traité de Maurolico donne une description complète des horloges solaires de toutes les espèces, suivant les différentes positions du plan de l'horloge. Les principes sont simples et uniformes, mais les cas se multiplient et remplissent les deux premiers livres. En fait, le livre premier commence en donnant toutes les notions préalables d'astronomie et de trigonométrie, avant de passer aux questions spécifiques des cercles et des lignes horaires. Le deuxième livre applique les coniques à les lignes horaires. et parvient à donner la *delineatio* ou description des courbes, en supposant connues la définition et les propriétés fondamentales.

---

11. On se réfère, bien entendu, au modèle astronomique de Ptolémée que Maurolico suit et défend contre Copernic.

12. Ces heures, considérées jusqu'au XVIIIe siècle, donnaient une division du temps, dans le jour ou dans la nuit, qui marquait expressément les moments à dédier aux travaux manuels ou, dans le cas des Babyloniens, aux observations d'astronomie.

3. Le troisième livre du *De lineis horariis* représente un véritable compendium des coniques, composé bien avant la parution du texte de Commandin en
1566. Comme a observé M. Clagett[13], Maurolico avance beaucoup sur les
anciens résumés de Giorgio Valla (1501) et de Johannes Werner (1523).

Maurolico commence avec les définitions et les premiers éléments en rassemblant les premières définitions d'Apollonius et les propositions 1-10 du
premier livre des coniques, notamment 3-5 et 7. Après ce chapitre préliminaire,
qui assure la fondation, il passe à la considération de la parabole, de l'ellipse
et de l'hyperbole en développant pour chacune, les propriétés fondamentales,
dans les chapitres I-III. Cela correspond aux propositions 11-13 (et 20) du livre
premier et 44-47 du livre deuxième d'Apollonius où l'on donne la construction
des diamètres et des axes. A partir de ces résultats il obtient les constructions
de ces courbes dans le plan.

Les chapitres suivants (IV-VII) sont bien plus intéressants dans la mesure où
ils traitent de problèmes plus difficiles : en effet, pour considérer les deux espèces de lignes horaires, on doit envisager les courbes comme données non seulement par leurs points mais aussi par leurs tangentes.

Maurolico commence, dans le chapitre IV, à traiter des sécantes et des tangentes en rassemblant de nombreuses propositions d'Apollonius, c'est-à-dire
les propositions 17, 18, 24, 25, 26, 27, 29 du livre premier et 13, 16, 33 du
livre deuxième, suivies par les propositions 20, 21, 35, 33, 36, 34, 37 du premier livre. Il observe que pour le dernier groupe des propositions (I, 33-37) il
a donné un ordre différent en modifiant les démonstrations avec l'appui de
deux lemmes, ce qui lui permet un approche plus directe en se référant toujours au cône dont on envisage les sections. Il dit explicitement de ces *quatuor
conclusiones* : *aliter, quam Apollonius, quod pulchrum fuit, ostendimus.*

Le chapitre V s'occupe de la doctrine des diamètres conjugués et de la construction des tangentes. Ici Maurolico réduit la théorie à des règles, qui
s'appuient quelques fois (I, 15, 16) sur des propositions d'Apollonius qui sont
citées en manière explicite, mais sont le plus souvent déduites des conclusions
du chapitre précédent.

Aussitôt suit le chapitre VI sur les asymptotes de l'hyperbole : comme
encore d'habitude à ce temps là, il parlait de *hyperbolae contrapositae* pour
envisager les deux branches de l'hyperbole des modernes. Maurolico s'étend
avec quelque longueur sur l'application aux lignes horaires des différents horloges, qui lui donnent des beaux exemples pour ces notions qui faisaient part
de la géométrie la plus avancée.

Enfin, le chapitre VII donne les théorèmes relatifs aux parallélogrammes
construits à partir des points sur l'hyperbole et sur ses asymptotes ; l'énoncé
figure expressément comme titre : *Quod parallelogramma inter Non tangentes*

13. M. Clagett, *Archimedes in the middle ages*, 4, Philadelphia, 1980, 331-334.

et periferiam locata, sunt invicem aequalia : quodque tam tangentis sectionem à tactu, quàm secantis eamdem à periferia ad Non tangentes, recepta segmenta sunt aequalia.

Ces trois conclusions, correspondent aux propositions 12, 8 et 3 du livre second d'Apollonius. Elles aussi, comme Maurolico souligne, *alio ordine modoque ostenduntur.*

4. On voit donc que Maurolico s'écarte de l'exposition apollonienne : en effet, après les éléments du début, il a constitué son traité par problèmes. Après avoir donné les constructions des trois coniques, aboutissant à la représentation plane qui aujourd'hui est canonique, il poursuit en donnant des règles déduites de théorèmes d'Apollonius qu'il ne démontre pas complètement.

C'est du reste son style quand il donne des *Compendia* : en partant des définitions il présente les raisonnements dans un ordre qui ne correspond pas toujours à l'ordre des propositions de l'auteur du texte qu'il va résumer ; après, il en développe les conséquences en esquissant tout simplement les lignes de la preuve. Mais il n'oublie pas d'aviser son lecteur quand il propose une marche qui diffère sensiblement de celle d'Apollonius. Il donne ses raisons : parfois il s'agit seulement de simplifier ; d'autres fois il exprime des considérations d'ordre méthodologique, par exemple quand il envisage directement le cône générateur des sections pour éviter des longues transformations des propriétés planimétriques et, encore, quand il substitue des démonstrations directes aux raisonnements par l'absurde[14].

On doit ici souligner qu'il donne une dérivation originale du latus rectum (qu'il appelle *recta diameter*) et que l'une des constructions planes de la parabole se rapporte, selon Clagett, à la tradition médiévale[15].

Ce petit traité de Maurolico a été bien apprécié dans le XVIIe siècle par tous ceux qui abordaient l'étude des sections coniques essayant de dépasser Apollonius. C'est ce que nous rappelle Borelli en disant : (…) *Fr. Maurolicus … lib. 3 de lineis horariis anno 1553 breviarium conicorum composuit, in quo egregias demonstrationes excogitavit linearum tangentium sectiones conicas et asymptotarum, quas ob eorum summam praestantiam duo viri praeclari Midorgius anno 1639, et Gregorius a S. Vincentio anno 1647 amplexati sunt, et iis sua opera exornarunt*[16].

Borelli se réfère ici aux propositions finales du chapitre IV dont nous avons parlé plus haut.

On voit donc l'importance de faire une analyse très détaillée de ce troisième livre de l'opuscule *De lineis horariis*, en le confrontant tant aux éditions clas-

14. Cela correspond aux discussions de ce temps là sur la logique et sur le statut des démonstrations mathématiques.

15. M. Clagett, *Archimedes in the middle ages, op. cit.*, 332.

16. G.A. Borelli, *Elementa conica Apollonii Pergei et Archimedis opera nova et breviori methodo demonstrata*, Romae, 1679, 2.

siques d'Apollonius qu'aux traités sur les coniques des " modernes ", notamment du XVIIe siècle, y compris Borelli.

Mais comme il s'agit d'un *Compendium* on peut faire aussi une comparaison interne à Maurolico même, c'est-à-dire entre sa lecture des classiques (en
ce cas Apollonius ex traditione Maurolyci) et son propre *Compendium*, qui,
comme nous avons montré, n'est pas tout à fait un simple résumé du texte
majeur. Cette comparaison nous montrera le progrès du mathématicien Maurolic dans la lecture des classiques de l'antiquité, et aussi quelle était son idée sur
le bases nécessaires pour apprendre les mathématiques aux commençants et les
faire progresser[17]. Avec l'édition complète de Maurolico que notre groupe se
propose d'accomplir on espère réaliser aussi cette tâche.

17. On peut faire une pareille comparaison avec Euclide, en se référant aux manuscrits. J'ai
commencé ce travail, qui se révèle fort intéressant. Mais le problème des *Compendia* de Maurolico
est plus vaste et requiert encore une grande attention.

A REFLECTION ON GALILEO'S EARLY EMPIRICAL SENSE

Thomas B. SETTLE

These thoughts are meant to contribute to the on going discussions dedicated to investigating Galileo's manuscript tract *De motu antiquiora*[1] (*ca.* 1590-1595). Among the aims have been identifying the sources of the materials in the manuscript itself, both written and otherwise, and coming to terms with Galileo's first major attempt to develop a new science of natural motion[2]. I take it that implicitly the investigations are proceeding on at least two levels. The first and obvious is the personal or biographical, accounting for Galileo and his work *per se* ; the second, and probably no less obvious, is the accounting for what we might call the Galileo phenomenon, the emergence in several locals of western Europe, in the late 16th and early 17th centuries, of early-modern, experimental natural philosophy. Galileo was only one of several in that period who, in retrospect, can be thought of as combining two activities : critically reviewing and eventually rejecting what was taken as Aristotelian natural philosophy and doing the work of turning certain proto-sciences of the Renaissance into what we would come to recognize as the early modern sciences.

What do I mean by the " proto-sciences " ? I have in mind such topics as hydrostatics and fluid flow, musical acoustics, resistance of beams to fracture, perspective, cartography, navigation, magnetic phenomena, anatomy and physiology, gnomonics, mechanics in the sense of the simple machines and

---

1. G. Galilei, *Le Opere di Galileo Galilei*, vol. II, a cura di A. Favaro, Edizione Nazionale, 20 vols ; I.E. Drabkin, " Galileo Galilei : On Motion ", in I.E. Drabkin and S. Drake (eds), *Galileo Galilei : " On Motion " and " On Mechanics "*, University of Wisconsin Press, 1960.

2. R. Giacomelli, *Galileo Galilei Giovane e il suo " De motu "*, Pisa, Domus Galileiana, 1949 ; T.B. Settle, " Galileo's Use of Experiment as a Tool of Investigation ", in E. McMullin (ed.), *Galileo : Man of Science*, Basic Books, 1967, 315-337 ; R. Fredette, " Galileo's *De motu antiquiora* ", *Physis*, 14 (1972), 321-348 ; J. Renn, P. Damerow, S. Rieger & M. Camerota, *Hunting the White Elephant. When and how did Galileo discover the law of fall ?*, Preprint 97, Berlin, Max Plank Institute for the History of Science, 1998 ; M. Camerota, M.O. Helbing, *Galileo's Early Dynamics in Context. Galileo's " De motu antiquiora " and the " Quaestiones de motu elementarorum " of the Pisan professors. A Preliminary Survey*, ms. in revision.

machinery and mechanics in the sense we would come to know as free or sys-
tematically constrained motion, ballistics, and percussion and quite probably
several others including, in my view, " material astronomy ". That is, not the
physical astronomy which could only come into existence with Galileo's tele-
scopic discoveries, but that astronomy which became more and more depen-
dent on the design and execution of precise and reliable measuring
instruments. In fact this period witnessed a prolonged effort in the inventing
and improving of all kinds of measuring and calculating instruments. Needless
to say, the growing mastery in all these areas of endeavor required increasingly
detailed attention to material phenomena of one type or another along with
efforts towards abstraction and reduction to rational systems usually, but not
always, mathematical.

In the case at hand, the *De motu antiquiora*, what at first glance might seem
a variant of a more-or-less typical, discursive text on a late Aristotelian theory
of motion (including hydrostatics, motion in free fall and on inclined planes
and projectile motion) proves to display a highly developed sense of precise
empirical investigation in the context of a concerted attempt to accommodate
a proposed theory of free fall to closely observed empirical results. In the
immediate sense, of course, the attempt failed, but the work fostered the investigations which led to results Galileo expressed both to Guidobaldo del Monte
in 1602, concerning pendular and related motions[3], and to Paolo Sarpi[4] in
1604, with the first written account (which survives, that is) of the law of free
fall. Eventually Galileo would provide much more elaborate, systematic
accounts of these motions in the *Discorsi*[5], over 30 years later, along with
detailed descriptions of how others might empirically verify some of his
claims. Here I allude both to the famous description of observing balls rolling
on inclined planes and to the more informal but none the less convincing
descriptions of pendular motions. But that is the end of the story.

The prelude, on the other hand, comes in the early 16th century, with respect
both to an empirical interest in the problem of free fall and to a growth or
maturing of what I would like to call an " experimental propensity " in investigating it. I have dealt with this subject elsewhere[6] ; suffice it to say here that
what was taken to be the Aristotelian position on the proportionality of weight
to speed in free fall was already being subjected to empirical test, and rejec-

3. G. Galilei, " Lettera a Guidobaldo del Monte ", *Le Opere di Galileo Galilei*, vol. X, 97-100
(29.XI.1602).

4. G. Galilei, "Lettera a Paolo Sarpi", *Le Opere di Galileo Galilei*, vol. X, 115-116
(15.X.1604).

5. G. Galilei, *Le Opere di Galileo Galilei*, vol. VIII, *op. cit.*

6. T.B. Settle, " Galileo and Early Experiment ", in R. Aris, *et al.* (eds), *Springs of Scientific
Creativity*, University of Minnesota Press, 1983, 3-20 ; C.B. Schmitt, " Experience and
Experiment : A Comparison of Zabarella's View with Galileo's in *De motu* ", *Studies in the
Renaissance*, vol. 16 (1969), 80-138 ; T.B. Settle, " Ostilio Ricci, a Bridge Between Alberti and
Galileo ", *Actes du XIIe Congrès International d'Histoire des Sciences*, Tome IIIB (1971), 121-126.

tion, in increasingly complex and detailed ways, before Galileo was born and while he was a young man. One culmination of these trends was in the work, including experimental, of Girolamo Borro, whose lectures in Pisa Galileo may have attended when he was a student at the university. In fact, one can reasonably see Galileo's own early *De motu* as an evolutionary development out of the work of Borro[7], Galileo taking up the empirical and theoretical investigation of free fall where Borro had left it, of course adding very important elements from other sources (most importantly, Archimedes). And on the evidence of the *De motu*, Galileo immediately superseded Borro in the complexity and imagination of his experimental design, in the precision and thoroughness of his observation and reporting of data, and in the seriousness and flexibility of his attempts to construct a mathematical armature which would render the empirical findings results of this early research coherent and rationally interconnected.

Nor was the *De motu* the only place in which Galileo exhibited these traits in the early days. The most obvious other locus was in his brief tract *La Bilancetta*[8], drafted even before he began the *De motu*. Without going into a lot of detail here, it is worth mentioning that the *Little Balance* is a marvel of experimental design. First, it works, should there be any doubters. Among other things, Galileo paid particular attention to how one might provide a precise, finely divided linear scale on one of the arms of the balance, and then to how the experimenter was to read the scale, the divisions being so fine as to make it difficult for them to be read by eye. For the first, Galileo proposed using a length of fine wire (presumably, in this case, a fine steel cord made for use on a musical instrument, such as a lute) and wrapping it very compactly around the arm, the diameter of the wire being the basic linear division. For the second, Galileo suggested that, instead of trying to count the divisions visually, one should use a pointed instrument such as a stiletto, drawing the point slowly across the wires. The investigator could then sense the " bumps " as the point passed over the wires, both in his fingers and by its sound, doing the counting from both tactile and audio feedback. As I have already mentioned, this works quite well. Galileo was only about 23 years old at the time.

Another subject, if less obvious because less documented and studied, in which Galileo displayed an early prowess for experimental-rational research, was in his work in collaboration with his father, Vincenzo Galilei[9], in what we

7. G. Borro, *De motu gravium et levium*, Firenze, 1575 ; T.B. Settle, " Galileo's Use of Experiment as a Tool of Investigation ", *op. cit.* ; M. Camerota & M.O. Helbing, *The Leaning Tower Experiment in Context, Borro, Buonamici and Galileo on falling bodies*, in publication.

8. G. Galilei, *Le Opere di Galileo Galilei*, vol. I, *op. cit.* ; L. Fermi and G. Bernardini, *Galileo and the Scientific Revolution,* New York, Basic Books, 1961, Appendix.

9. C.V. Palisca, 1961, " Scientific Empiricism in Musical Thought ", in H.H. Rhys (ed.), *Seventeenth Century Science and the Arts*, Princeton U. P., 1961, 91-137 (v. 120-137) ; C.V. Palisca, *The Florentine Camerata : Documentary Studies and Translations*, Yale University Press, 1989.

might call musical acoustics. The story is complex and there are yet many facets of it to clarify, but let me try to summarize what I think is legitimate to say as of now. One aspect of it has to do with the young Galileo's discovery of the isochronous properties of the simple pendulum. Traditionally we have supposed that this happened while he was still a student at the University of Pisa, in 1585 or somewhat before[10]. So far as I am concerned that may well be the case, but he could as well have made the discovery during the four years between 1585 and 1589, when he was home, in Florence, in part collaborating with his father Vincenzo, before returning to Pisa as a professor of mathematics.

In what did his collaboration with his father consist ? We know that Vincenzo was a practising musician, a teacher of music, a composer of music and a writer on the theory of music[11]. At one point he had the opportunity of studying with Gioseffo Zarlino, the author of what, for simplicity, I will call a number-theory approach purporting to explain the reasons or causes for harmonic ratios. For Zarlino, two musical tones were in harmony when is some sense they exhibited " ratios " using the first six natural numbers, one to six (hence the name attached to the theory, the senario). While the theory in principal was supposed not to have anything to do with the material reality of sounds or instruments, there was, nevertheless, an implicit reference to the tones exhibited on a monocord, or on any stringed instrument, such as a lute. A string of a given length, when plucked, sounds a tone. When stopped in the middle and plucked again, it sounds a note an octave higher than the first. The ratio essentially characterizing the octave is two-to-one, the same as the ratio of the lengths of the string.

Now for a while Vincenzo Galilei accepted these ideas ; but by the mid-1580s he had come to doubt them and, what is more important, to reject them on the basis of his own experimental research. Given that the ratios work well enough when referred to a monocord, they fail completely when some variable other than the lengths of a string is used. For instance, if we take a convenient, steel lute-string, hang it from an overhead beam and attach different weights to its lower end, we find that the ratio of weights needed to produce an octave is not two-to-one, but four-to-one[12]. In sum, there is no unique set of numbers that can be assigned to the several harmonic ratios. For Vincenzo, this meant the end of Zarlino's senario and the end of any theoretical underpinnings for musical harmonies. The rules for composing music and tuning instruments were to be based on the experience and practice of musicians, i.e., on the

10. V. Viviani, *Questa del pendolo...*, BNCF, s.d., Mss. Galileiana, 227, c. 60r ; trascrizione : R. Caverni, *Storia del metodo sperimentale in Italia*, vol. 1, 1891-1900, 303, 6 vols ; V. Viviani, " Raccconto istorico della vita di Sig. Galileo Galilei ", *Le Opere di Galileo Galilei*, vol. XIX (1654), 597-632 ; V. Viviani, " Lettera... al Principe Leopoldo... intorno all'applicazione del pendolo all'orologio ", *Le Opere di Galileo Galilei*, vol. XIX (1659), 647-659.

11. V. Galilei, *Dialogo della musica antica et della moderna*, Firenze, 1581.

trained ear. For Galileo, it meant the beginning of a life-long search for another basis for a theory, or science, of musical harmonics. And it gave him an important clue, a significant starting point for his further research.

In fact, a steel lute-string, hanging from an overhead beam, is two things. First, as indicated, it is a means of experimenting with musical tones. In doing so one plucks the cord and listens to the tone produced. One can also see the blur of the actual vibrations of the cord. If one is paying attention, one will both hear the strength or loudness of the sound diminish and see the amplitude of vibration of the cord diminish with it, while the tone itself remains the same[13]. This must have been a commonplace, known from antiquity.

But in his immediate context Galileo must have been struck by a strong analogy with what he had found or was finding with pendulums. The same lute string with a weight on its end is, of course, a pendulum. When it is set swinging, the amplitude of its swing will also slowly diminish, while the frequency of swing remains constant[14]. In the immediate context, this suggested to Galileo that the distinguishing feature of a musical tone was the rate of vibration of the original source of the tone and the subsequent vibrations or pulses in the air transmitting it. I said " suggested " ; Galileo was clearly aware that an analogy was not a proof. His account of pendular motions in the *Discorsi* are strong evidence for this caution ; but he did continue to search for direct evidence of the relationship of frequency to tone. The search yielded results which he felt supported the idea, but which he also knew would not be entirely convincing to others.

Nevertheless, this early discovery, probably in the period 1585-1589, set much of the style for his later work. The whole story is much too long to recount here ; let me just mention a few strands. First, what he had found was another category of " natural motion ", a category not treated in any version of the traditional theories of motion, but which included those motions exhibited by what might be called " natural oscillators ", objects or systems which, once set vibrating, continue to do so with a constant frequency. These would include gongs and bells as well as vibrating strings and simple pendulums. And even-

---

12. V. Galilei, *Discorso particolare intorno alla diversita delle forme del diapason*, BNCF, Mss. Galileiana, 3, ff. 44r-54v ; trascrizione e traduzione in inglese : C.V. Palisca, 1989, 180-197 ; C.V. Palisca, " Was Galileo's Father and Experimental Scientist ? ", *Music and Science in the Age of Galileo,* in V. Coelho (ed.), Dordrecht, Boston, Kluwer, 1992, 143-151 (University of Western Ontario series in philosophy of science, v. 51) ; P. Gozza (a cura di), *La musica nella rivoluzione scientifica del seicento*, Il Mulino, 1989.

13. T.B. Settle, " La rete degli esperimenti galileiani ", in M. Baldo Ceolin (a cura di), *Galileo e la scienza sperimentale*, Padova, 1995 ; T.B. Settle, " Galileo's Experimental Research, an Experimental Approach " (the English version of Settle, 1995), in T.B. Settle (ed.), *Galileo's Experimental Research*, Preprint 52, Berlin, Max Planck Institute for the History of Science, 1996, 5-37.

14. T.B. Settle, " The Pendulum and Galileo ", in T.B. Settle (ed.), *Galileo's Experimental Research*, Preprint 52, Berlin, Max Planck Institute for the History of Science, 1996, 39-49.

tually one such natural oscillator would become a central feature in his attempt to offer a proof of the daily and annual motions of the earth. I am referring to the tides, sea tides[15]. As we know, he originally intended calling his *Dialogo ... sopra i due massimi sistemi ...* something like a *Dialogue on the Ebb and Flow of the Sea*, and in the Fourth Day of that work he presented his chief, earthly evidence for the physical reality of Copernicanism. He had observed that when a container filled with water is jolted from the outside, the water on the inside will rush back and forth, piling up first at one end then at the other. And he found to his satisfaction that this was another example of a naturally oscillating system, the period depending on the characteristics of the container : length, width and depth. For Galileo, in sufficiently large sea or ocean basins the water was set and kept naturally oscillating by the impulses resulting from a combination of the daily and annual motions of the earth[16] ; and, there being no other mechanical cause for tidal phenomena, their mere existence argued for that Copernican double mobility. And there is fair reason to believe that he had already begun formulating these ideas as early as 1597, and quite possibly earlier.

I would suspect, then, is that while he was working on what we have come to know as the *De motu antiquiora*, he already had in mind including a chapter or chapters on such natural oscillators. As we know, he ceased drafting the *De motu*, in all likelihood, when he realized that the several initial assumptions for that work were no longer tenable.

I have suggested elsewhere that they became untenable for him as a result of his continued attempts to match empirical results to those assumptions. My view is that he perfected what we have come to know as the inclined plane experiment, the one described in the Third Day of the *Discorsi*, while still in Pisa, or in the first few years at Padua. Now, a key element in that experiment was his so-called water clock, or better a " water timer ". Let us ask ourselves : how was it that Galileo could have had confidence that this water timer gave reliable and precise results ? Without any positive statement on the part of Galileo, my guess is that the only way he could have gained such confidence would have been by checking it against a simple pendulum, which he was already the master of.

For Galileo, then, the pendulum was of key importance to the whole edifice of his attempt to construct a new science of several natural motions, a new physics, even if — and this must have been a great source of disappointment

15. G. Borro, *Dialogo del flusso e reflusso del mare*, Lucca, 1561 ; G. Borro, *Del flusso e reflusso del mare & dell'inondatione del Nilo : la terza volta ricorretto dal proprio autore*, Firenze, 1583 ; P. Ventrice, *La Discussione sulle Maree tra Astronomia, Meccanica e Filosophia nella Cultura Veneto-Padovana del Cinquecento*, Memorie, Classe di Scienze Fisiche, Matematiche e Naturali, Istituto Veneto di Scienze, Lettere ed Arti, vol. XXXIV, Fasc. III (1989).

16. N. Swerdlow, *The Fourth Day : The Cause of the Tides*, 1997, unpublished ms.

to him — he never succeeded in rationally connecting pendular motion with linear free fall, much less in " resolving " percussion.

All this said, there was another, shall we say " more dramatic " aspect to his early musical and pendular discoveries. What he learned both from what his father had done and from his own labors was that the most elaborate and convincing theories could be thoroughly rejected on the basis of material, experimental evidence and that the same experimental evidence could yield truths about natural phenomena, unpredicted and even unimagined in ancient or even contemporary theory. As we know, finding things unimagined by the ancients became more-or-less a habit with him.

It would seem clear, then, that already in his early 20's Galileo had developed very strong experimental sensibilities and capacities. Even before his return to Pisa in 1589 he had discovered novelties and had embarked on projects which would require him to revise accepted theory and even to construct or invent new mathematical structures for what he was seeing empirically, even to the point of contributing to mathematics itself. But clearly there is more eventually to be said about the beginnings of these empirical sensibilities. We may ask, for instance, why Girolamo Borro was led to put both Aristotle and his own Aristotelianism to the test by conducting his own dropping experiments ? Or why was it that Vincenzo Galilei felt authorized, on the basis of " mere material results " to deny what had become a widely accepted theory of musical harmonies ? What was in this acceptance, on both their parts, of the primacy of empirical data over previously developed theoretical positions ?

MOTION ON INCLINED PLANES AND IN LIQUIDS IN GALILEO'S EARLIER *DE MOTU*

Pierre SOUFFRIN

The purpose of this study[1] is to reconsider Galileo's treatment of motion on inclined planes in his earliest written works devoted to motion, known as *De motu antiquiora* (hereafter *DMA*), avoiding the usual clear-cut separation with the treatment given, in the same work, to the problem of motion in media. It is believed that some clarifications arise from this unconventional approach concerning the relations between the young Galileo's views on dynamics and his terminology on the one hand, and the scholastic tradition on the other hand.

THE CASE FOR MOTION ON INCLINED PLANES IN *DMA*

In the 14th chapter of the version in 23 chapters of *DMA in quo agitur de proportionibus motuum eiusdem mobilis super diversa plana inclinata*, Galileo considers two questions (here and in the following (N, M ; ch. P) stands for *Opere*, vol. I, from page N line M onwards, chapter P)[2] : (296, 12 ; ch. 14) *Quaeritur enim cur idem mobile grave, naturaliter descendens per plana ad planum horizontis inclinata, in illis facilius et celerius movetur quae cum hor-*

---

1. This is part of a series devoted to the history of the theory of motion including " Galilée et la tradition cinématique pré-classique, la proportionnalité velocitas-momentum revisitée ", *Cahier du Séminaire d'Epistémologie et d'Histoire des Sciences*, n° 22 (Nice, 1990), 89-123 ; " Sur l'histoire du concept de vitesse – Galilée et la tradition Scolastique ", in B. Ribémont (sous la direction de), *Le temps, sa mesure et sa perception au Moyen Age*, Caen, Paradigme, 1992, 243-268 ; " Galilée, Torricelli et la " loi fondamentale de la dynamique scolastique ". La proportionalité *velocitas-momentum* revisitée ", *Sciences et Techniques en Perspective*, 25 (1993), 122-134. These papers can be read also on the web at http://wwwrc.obs-azur.fr/ cerga/ hdsn/ Psouffrin/ souffrin.html. Also relevant are " Sur l'histoire du concept de vitesse d'Aristote à Galilée ", *Rev. Hist. Sci.*, 45 (1992), 231-267 and J.L. Gautero, P. Souffrin, " Note sur la démonstration " mécanique " de […] l'isochronisme des cordes du cercle dans les *Discorsi* de Galilée ", *Rev. Hist. Sci.*, 45 (1992), 269-280.

2. Galileo Galilei, *Le opere*, Firenze, Barbera, 1899-1909, reprinted 1964-1968 and 1968.

izonte angulos recto propinquiores continebunt ; et, insuper, petitur proportio talium motuum in diversis inclinationibus factorum.

He claims that he has succeeded in solving these problems, *ex notis et manifestis naturae principis.*

His answer to the second question, viz. to the quantitative problem, is first given in the form : (298, 23 ; ch. 14) *Eandem ergo proportionem habebit celeritas in* ef *ad celeritatem in* gh, *quam linea* da *ad lineam* pa. *Est autem sicut* da *ad* ap *ita* qs *ad* sp, *hoc est obliquus descensus ad rectum descensum.*

He gives the following figure :

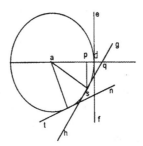

Here the motion on an inclined plane is compared to the vertical free fall over the same difference in height. This is easily transformed into the comparison of the motions on two inclined planes with different slopes, answering precisely the problem as it is stated : (301, 23 ; ch. 14) *Constat ergo, eiusdem mobilis in diversis inclinationibus celeritates esse inter se permutatim sicut obliquorum descensuum, aequales rectos descensus compraehendentium, longitudines.*

THE CONVENTIONAL ANALYSIS OF THIS THEORY OF MOTION
ON INCLINED PLANES

This answer, as well as its derivation, have both been considered faulty by modern critics.

The proposition (301, 23) itself is considered to be grossly wrong because it is supposed to imply, in its very wording, the uniformity of the motion on an inclined plane[3]. I did challenge this conclusion elsewhere[4] ; this led me to see

---

3. E.g. P. Galluzzi, *Momento*, Roma, Edizione dell'Ateneo & Bizzarri, 1979, 182 : *E inutile sottolineare ... che la teoria galileiana del De motu, la " dinamica pisana ", come è stata definita, è una dinamica dei moti uniformi* ; S. Drake, *Galileo at Work*, Chicago, London, 1978, 24 : " except for its treatment of speeds on inclined planes, chapter 14 [of the *DMA*] was sound " ; W.L. Wisan, " The New Science of Motion : A Study of Galileo's *De motu locali* ", *AHES*, 13 (1974), 103-306. One reads on p. 150 : " Galileo assumes the motion along these lines to be uniform ".

in this *de motu theorem* nothing but an earlier formulation of the isochronism of the descent along the cords of a circle the final form of which is the " Theorem VI " on accelerated motion in the *Discorsi*. This question — the meaning of the proposition — will not be considered further here.

The lengthy demonstration which Galileo provides for his Proposition is also claimed by the modern critics to be unsatisfying, either for a lack of well defined concept of *celeritas*[5] or for the use of an unjustified proportionality (*proportionalitas*)[6].

Such analyses by distinguished scholars of course reflect the considerable complexity of the conceptual and the terminological aspects of the text, and one finds noticeable variations of interpretation between authors. It is clear that a definite understanding of this text implies definite interpretations of such crucial terms as *celeritas, tarditas, facilitas* and of different occurrences of *vis* in the text (e.g. ... *tanto maiori vi fertur ;* ... *descendat, quanto minori vi trahi sursum,* ...).

It is worthwhile noticing that these alleged weaknesses in Galileo's text have not hitherto been considered to arise in his discussion of natural motion in media, which constitutes the main part of the treatise. If I can show that both theories are but two instances of a unique theory, then we shall be confronted with two possibilities, either that some of the flaws found in the theory of the inclined plane must also occur in the theory motions in media, or that the " flaws " are artefacts due to a misunderstanding of Galileo's text. In both cases some new insight should arise from a detailed comparison of these two theories as they are expounded in *DMA*. I now consider such a comparison.

I must be emphasized that I shall restrict the comparison in this study to Galileo's analysis of the motion of " one and the same body " in different situations, although the full problem includes of course the general case of different bodies in different situations. Besides the conditions set by the limited format of this contribution, there is an historical justification for a separate presentation of this restricted problem : the full problem is actually solved by Galileo in the *DMA* for the case of motions in media, but not for the case of motions on inclined planes. For this latter case Galileo contents himself with relying on the skill of his reader to make such generalizations : *haec et similia ab his, qui quae supra dicta sunt intellexerint, facile inveneri possunt.* How-

4. See B. Ribémont (sous la direction de), *Le temps, sa mesure et sa perception au Moyen Age, op. cit.* ; " Velocitas totalis. Enquête sur une pseudodénomination médiévale ", in S. Caroti, P. Souffrin (eds), *La nouvelle physique du XIVe siècle*, Firenze, Olschki, 1987, 251-276. (available on the web).

5. E.g. I.E. Drabkin in S. Drake, I.E. Drabkin, *Galileo Galilei On Motion and On Mechanics*, Madison, 1960, 8 : " ... the proof erroneously links this force with the still unclarified notion of the speed of free fall ".

6. E.g. P. Galluzzi, *Momento, op. cit.*, 195 : ... *Stabilite le proporzioni della " gravità " del mobile lungo piani diversamente inclinati, Galileo ritiene di poterle direttamente estendere alle velocità* (celeritates).

ever, the generalization consistent with the limited problem as solved by Galileo is far from evident and raises interesting problems which will be considered in a forthcoming paper.

THE THEORY OF MOTION IN *DMA* RECONSIDERED

Motions of a body in media and on inclined planes

Let us first consider the logical structure of Galileo's demonstrations of his propositions concerning motion in media (chapters 7 ; 8 ; 9) and on inclined planes (chapter 14). I claim that close analysis shows that both situations are analyzed within a unique demonstrative scheme. To support this claim I shall in the following consider in parallel the successive steps of the arguments as they appear in the *DMA*, and give for each of the two problems the relevant (or representative) passages.

1.- The problem is first set, in similar terms, as the comparison of the motions of one and the same body in different media, or on different planes.

a) in media

(260, 6 ; ch. 7) *Cum in superioribus satis abunde explicatum sit, quomodo motus naturales proveniant a gravitate et levitate, nunc videndum est unde accidat maior aut minor ipsius motus celeritas.*

b) on planes

(296, 12 ; ch. 14) *Quaeritur enim cur idem mobile grave, naturaliter descendens per plana ad planum horizontis inclinata, in illis facilius et celerius movetur quae cum horizonte angulos recto propinquiores continebunt ; et, insuper, petitur proportio talium motuum in diversis inclinationibus factorum.*

2.- The causes and effects are identified and their connection asserted.

2.1.- The causes of the *celeritates* are the same as the causes of the motions, since motion and *celeritas* are one and the same thing.

a) in media

(260, 6 ; ch. 12) *In utroque motu ex eadem causa pendere tarditatem et celeritatem, nempe ex maiori vel minori gravitate mediorum et mobilium, mox demonstrabimus ;*

(261, 17 ; ch. 7) *ut veram tarditatis et celeritatis motus causam afferamus, attendendum est, celeritatem non distingui a motu : qui enim ponit motum, ponit necessario celeritatem ; et tarditas nihil aliud est quam minor celeritas. A quo igitur provenit motus, ab eodem provenit etiam celeritas : cum itaque a gravitate et levitate motus proveniat, ab eadem ut tarditas vel celeritas proveniant, necessarium est ;*

b) on planes

(261, 17 ; ch. 7) also applies, except for the reference to *levitas*.

2.2.- Since the effects are as the causes, the *celeritates* are as (*i.e.* proportional to) the causes of the motions.

a) in media (downwards)

(262, 3 ; ch. 7) *Quare manifestum est, quod, si invenerimus in quibus mediis idem mobile gravius extiterit, inventa erunt media in quibus citius descendet ; quod si, rursus, demonstremus, quantum idem mobile gravius sit in hoc medio quam in illo, erit, rursus, demonstratum, quanto citius in hoc quam in illo deorsum movebitur ...*

b) on planes

(297, 6 ; ch. 14) *Si itaque inveniamus quanto minori vi trahitur sursum grave per lineam bd quam per lineam ba, erit iam inventum quanto maiori vi descendat idem grave per lineam ab quam per lineam bd et, similiter, si inveniamus quanto maior vis requiritur ad sursum impellendum mobile per lineam bd quam per be, erit iam compertum quanto maiori vi descendet per bd quam per be.*

3.- The measure (or ratio[7]) of the causes of the motions is investigated.

3.1.- As a general statement, this measure (or ratio) is claimed to be equal to the measure (or ratio) of the forces necessary to prevent the motion.

a) in media (downwards)

(275, 1 ; ch. 9) *Querimus igitur, sphera plumbea quanta vi deorsum fertur in aqua. Patet igitur, primo, quod sphera plumbea fertur tanta vi, quanta requeritur ad illam sursum attrahendam.*

[A similar statement is found at (270, 3 ; ch. 8), and the equivalent one for motion upwards at (274, 17 ; ch. 9)]

b) on planes

(297, 2 ; ch. 14) *Ut igitur haec consequi possimus, prius hoc est considerandum, quod etiam supra animadvertimus : scilicet, quod manifestum est, grave deorsum ferri tanta vi, quanta esset necessaria ad illud sursum trahendum ; hoc est, fertur deorsum tanta vi, quanta resistit ne ascendat.*

3.2.- This force (or resistance) able to prevent the motion is claimed to be equal to the *gravitas* accounting for the particular situation considered, *i.e.* to the *gravitas in loco*.

a) in media (downwards)

(271, 16 ; ch. 8) *Restat igitur ut [...] ostendamus proportionem quam servant celeritates eiusdem mobilis in diversis mediis : quae omnia ex hac demonstratione facile haurientur. Dico igitur, solidam magnitudinem aqua graviorem deorsum ferri tanta vi quanto aqua, molem habens aequalem moli ipsius magnitudinis, levior est ipsa magnitudine.*

---

7. In preclassical natural philosophy the concept of measure of quantities is identified with the concept of ratio (*proportio*) of quantities of the same species. It is too often foreseen that this situation will change only with the — later — definition and use of derived quantities.

[and (269, 25 ; ch. 8) for motions upwards]

b) on planes

(297, 12 ; ch. 14) *Sed tunc sciemus quanto minor vis requiratur ad sursum trahendum mobile per bd quam per be, quando cognoverimus quanto eiusdem mobilis maior erit gravitas in plano secundum lineam bd, quam in plano secundum lineam be.*

4.- Inserting these results in the proportionality stated above in (2.2) solves the problem.

a) in media (downwards)

(272, 20 ; ch. 8) *Hac igitur demonstratione percepta, quaestionum exitus facile dignosci potest. Constat enim, idem mobile in diversis mediis descendens eam, in suorum motuum celeritate, servare proportionem, quam habent inter se excessus quibus gravitas sua mediorum gravitates excedit ...*

a') in media (upwards)

(270, 29 ; ch. 8) *Patet igitur, universaliter, celeritates inter se motuum sursum, esse, sicut excessus gravitatis unius medii super gravitatem mobilis se habet ad excessum gravitatis alterius medii super gravitatem eiusdem mobilis.*

b) on planes

(298, 26 ; ch. 14) *... constat igitur, tanto minori vi trahi sursum idem pondus per inclinatum ascensum quam per rectum, quanto rectus ascensus minor est obliquo ; et, consequenter, tanto maiori vi descendere idem grave per rectum descensum quam per inclinatum, quanto maior est inclinatus descensus quam rectus.*

In this last quotation, the final sentence is the solution for the inclined plane problem.

A traditional demonstrative scheme

Both problems are seen to be treated in a fully consistent way along the following demonstrative scheme :

A dynamical rule — or law — is first accepted in the generic form of the proportionality (*proportionalitas*) between causes and effects (item 2.2), and is then transformed into a quantitative form by introducing the adopted measures for cause and effect (in item 3 above for the cause, and according to the pre-classical tradition for the *celeritas*[8] designated as the effect).

This general dynamical statement is nothing but a variant of the standard dynamical rule discussed by scholastic commentators, stating that the *velocitas* is proportional to the *potentia* and inversely proportional to the *resistentia*[9]. In the restricted problem considered here of the motions of one and the same

---

8. On this tradition, see J.L. Gautero, P. Souffrin, " Note sur la démonstration " mécanique " de [...] l'isochronisme des cordes du cercle dans les *Discorsi* de Galilée ", *op. cit.* ; and " Sur l'histoire du concept de vitesse d'Aristote à Galilée ", *Rev. Hist. Sci.*, XLV (1992), 231-267.

body the *resistentia* plays no role, but I shall just mention that a concept of *resistentia* is actually implied and used by Galileo in the DMA in a quantitative way, although not explicitly, when he proves that a large piece of a given material rises in a heavier media with the same velocity as a larger piece of the same material, or when he let his reader extend his analysis to the case of the fall along different inclined planes of bodies of different *gravitates*[10].

Some consequences of the hypothesis of internal consistency of DMA chapter 14 on the interpretation of Galileo's scientific terminology

This analysis, together with the *consistency hypothesis* that Galileo's demonstrations and discussions in this chapter 14 of the DMA are logically deductive rather than somewhat metaphoric at critical places, allows for a clarification of some terminological problems. The consistency hypothesis is accepted here but is by no mean trivial ; in my view, the stronger argument for it is precisely the unity of Galileo's dynamical discussions pointed to above.

Consider first two crucial occurrences of vis. The " consistency hypothesis " implies that in item (4, b), *i.e.* (296, 26 ; ch. 14), *tanto minori vi trahi sursum idem pondus* the term *vis* has a dynamical meaning referring to the cause of the motion, rather than a kinematical meaning referring to the effect[11], while in the same sentence *tanto maiori vi descendere* must be understood in the kinematical meaning of " descends with a velocity so much larger ".

More difficult is a correct interpretation of Galileo's use of *facilitas* which occurs in the DMA at least five times in expressions such as *facilius* et *citius*, as hereabove in item 1.

We may first notice that the statement of the problem in item 1 above, as *quaeritur ... cur ... facilius et celerius movetur ...* does actually introduce the two questions investigated in the text following, specifically the variations of the gravitas in plano and of the celeritas with the slope of the planes if and only if *facilitas* refers to the cause and *celeritas* to the effect.

The same conclusion arises with still more strength when applying the *consistency hypothesis* to the much debated section of DMA chapter 14 where Galileo puts together his arguments before concluding to the proportionality between the *celeritates* and, say, the slopes of the planes. The text goes as follow (where I emphasize the terms discussed, and the figure is as above) :

9. For a documented presentation of this dynamics, see E. Grant, *Physical Science in the Middle Ages*, J. Willey, 1971 ; Cambridge, 1977.

10. For motions in media : (263, 14; ch. 8) ; for inclined planes : (301, 35; ch. 14).

11. As in the in Commandino's translation of Archimedes Book I " On floating bodies ", Proposition VI, which seems to be the relevant reference here : *Solidae magnitudines humido leviores, in humidum impulsae sursum feruntur tanta vi, quanto humidum molem habens magnitudini aequalem, gravius est ipsa magnitudine.* See M. Clagett, *Archimedes in the Middle Ages*, vol. 3, part III, Philadelphia, 1978, 642.

(298, 16 ; ch. 14) *Sed quanto maiori vi moveatur per ef quam per gh, ita innotescet : extensa, scilicet, linea ad extra circulum, quae secet lineam gh in puncto q. Et quia tanto facilius descendit mobile per lineam ef quam per gh, quanto gravius est in puncto d quam in puncto s ; est autem tanto gravius in puncto d quam in s, quanto longior est linea da quam linea ap ; ergo mobile eo facilius descendet per lineam ef quam per gh, quo linea da longior est ipsa pa. Eandem ergo proportionem habebit celeritas in ef ad celeritatem in gh, quam linea da ad lineam pa.*

If we are to recognize in this short section a synthetic restatement of the arguments developed at length in the preceding discussion, then *facilitas* must be understood here as the denomination of the cause of the motion ; it is then what scholastics used to call inclinatio, and what Galileo himself in the tract *Le meccaniche* written soon after called *propensione* or *momento*.

CONCLUSIONS

According to the above analysis Galileo's demonstration of the inclined plane theorem in *DMA* is devoid of the ambiguities which modern critics find in it.

An essential implication of my interpretation concerns Galileo's scientific terminology. More specifically, it implies the understanding of the term *facilitas* as it appears in the *DMA* is as a cause of motion, as an evolution of the term inclinatio of scholastic dynamics. And the latter is, in my opinion, came to have the same meaning as the propensione or momento which Galileo was soon to use in *Le meccaniche* and further works. This interpretation accords with my previous claim that chapter 14 of the *DMA* belongs to what I called the *velocitas-momentum* class of Galileo's texts[12].

The hitherto unnoticed unity of Galileo's treatment of the motion of a body on different inclined planes and in different media which the present paper demonstrated, means that these two physical problems are treated in *DMA* with one and the same theory of motion, which is nothing but a variant — however important it may be — of the so-called traditional " scholastic dynamics ".

---

12. See " Galilée et la tradition cinématique pré-classique, la proportionnalité *velocitas-momentum* revisitée ", *op. cit.*, Part VI : " La classe des textes *velocitas-momentum* chez Galilée ".

PROJECTILE TRAJECTORY AND HANGING CHAIN — CHALLENGING THE NAIVE VIEW OF SCIENTIFIC DISCOVERIES

Peter DAMEROW, Jürgen RENN, Simone RIEGER

In the commonly accepted procedure of reconstructing discoveries in Galileo studies, the emphasis is mostly on finding clues for dating his discoveries. The interpretation of such clues presupposes, of course, a cocksure answer to the question of how a discovery can be identified. In the following a couple of findings will be presented, resulting from scrutinizing the sources, which may well contribute to the dating of Galileo's discoveries. This, however, is not our primary concern. What we want to show is that the answer to the questions concerning the nature of discoveries is not at all obvious. It may well be that premature answers to these questions have, as will also become clear, led historians of science astray. We argue that the widely accepted datings[1] of Galileo's discoveries of the law of fall to 1604 and of the parabolic trajectory to around 1607/1608 result from a too simple-minded idea of discovery and a highly selective reading of the sources.

THE NEGLECTED ISSUE

Although the discoveries of the law of fall and of the parabolic trajectory are among the most investigated issues in Galileo studies, it has nearly been completely neglected that, for Galileo, there is a close connection between the parabolic trajectory and the catenary, that is, the curve of a hanging chain. Even in his published works Galileo emphasized the close connection between both objects of his research. In the *Discorsi* Galileo claims (through his spokesman Salviati) to have developed two methods for drawing parabolas.

---

1. *Cf.*, e.g., A. Koyré, *Etudes Galiléennes*, Paris, Hermann, 1966, 83 ; F. Klemm, " Der junge Galilei und seine Schriften De motu und Le mecaniche ", *Sonne steh still : 400 Jahre Galileo Galilei*, Mosbach, Physik Verlag, 1964, 68-81, 79-80 ; G. Galilei, *Two New Sciences*, Madison, University of Wisconsin Press, 1974, IX ; S. Drake, *Galileo : Pioneer Scientist*, Toronto, 1990, XIIIf. ; *cf.* also A. Van Helden, E. Burr, http://es.rice.edu/ES/humsoc/Galileo/galileo timeline.html (version of August 5, 1996).

" There are many ways of drawing such lines, of which two are speedier than the rest ; I shall tell these to you. One is really marvellous, for by this method, in less time than someone else can draw finely with a compass on paper four or six circles of different sizes, I can draw thirty or forty parabolic lines no less fine, exact, and neat than the circumferences of those circles "[2].

The first of these methods uses the path of a projectile : " I use an exquisitely round bronze ball, no larger than a nut ; this is rolled on a metal mirror held not vertically but somewhat tilted, so that the ball in motion runs over it and presses it lightly. In moving, it leaves a parabolic line, very thin, and smoothly traced. This [*parabola*] will be wider or narrower, according to the ball being rolled higher or lower. From this, we have a clear and sensible experience that the motion of projectiles is made along parabolic lines, an effect first observed by our friend [*i.e.* Galileo], who also gives a demonstration of it. We shall all see this in his book on motion at the next meeting. To describe parabolas in this way, the ball must be somewhat warmed and moistened by manipulating it in the hand, so that the traces it will leave shall be more apparent on the mirror "[3].

Galileo's second method uses a hanging chain : " The other way to draw on the prism the line we seek is to fix two nails in a wall in a horizontal line, separated by double the width of the rectangle in which we wish to draw the semiparabola. From these two nails hang a fine chain, of such length that its curve will extend over the length of the prism. This chain curves in a parabolic shape, so that if we mark points on the wall along the path of the chain, we shall have drawn a full parabola. By means of a perpendicular hung from the center between the two nails, this will be divided into equal parts "[4].

Near the end of the *Discorsi* Galileo furthermore stressed the close dynamical connection between the trajectory of projectile motion and the curve of a hanging chain : " Well, Sagredo, in this matter of the rope, you may cease to marvel at the strangeness of the effect, since you have a proof of it ; and if we consider well, perhaps we shall find some relation between this event of the rope and that of the projectile [fired horizontally]. The curvature of the line of the horizontal projectile seems to derive from two forces, of which one (that of the projector) drives it horizontally, while the other (that of its own heaviness) draws it straight down. In drawing the rope, there is [likewise] the force of that which pulls it horizontally, and also that of the force of the weight of the rope itself, which naturally inclines it downward. So these two kinds of events are very similar "[5].

2. G. Galilei, *Two New Sciences*, op. cit., 142.

3. G. Galilei, *Two New Sciences*, op. cit., 142f.

4. G. Galilei, *Two New Sciences*, op. cit., 143.

5. G. Galilei, *Two New Sciences*, op. cit., 256.

The sources that survived the erosion of time, and in particular Galileo's famous notes on motion preserved at the *Biblioteca Nazionale Centrale* in Florence as Codex Ms. Gal. 72[6], provide evidence that these remarks in the *Discorsi* correspond to real experiments performed in the course of Galileo's discoveries of the law of fall and of the parabolic trajectory.

This is obvious in the case of the *second* method. Among the pages of Ms. Gal. 72 we find a sheet[7] which shows that Galileo had used a chain for drawing catenaries of different angles, as well as a sheet[8] in which he used these catenaries as templates for constructing parabolic trajectories of different inclinations[9].

Concerning the *first* method the evidence is somewhat more indirect. At the end of a notebook of Guidobaldo del Monte[10], one of the closest collaborators and correspondents during Galileo's early research on mechanics, one finds the protocol of an experiment which is perfectly resembled by the description in the *Discorsi*[11]. This protocol not only describes precisely the experimental setting but also the symmetry of the resulting curve and its connection to the curve of a hanging chain, resulting from the same configuration of forces. Given that Galileo was present — that is what we are about to show — he must have immediately recognized that the outcome of this experiment is incompatible with the traditional view, that the trajectory consists at least theoretically of a straight line in the direction of the shot, followed by a curved line which is part of a circle ending in a perpendicular line representing the fall of the projectile downwards. It is well known that also Galileo himself had still taken for granted this theory in his early treatise *De Motu*, that is, about 1590.

DATING OF THE DISCOVERY

The neglected issue of the link between projectile trajectory and catenary represents a challenge also for the standard dating of Galileo's discoveries. Is

6. Ms. Gal. 72 in the Biblioteca Nazionale Centrale, Florence ; *cf.* the electronic representation of the manuscript in the internet : *Galileo Galilei's Notes on Motion : An Electronic Representation of the Manuscript Ms. Gal. 72* (1998), http://www.mpiwg-berlin.mpg.de/ Galileo Prototype/MAIN.html.

7. *Galileo Galilei's Notes on Motion : An Electronic Representation of the Manuscript Ms. Gal. 72, op. cit.,* folio 41/42.

8. *Galileo Galilei's Notes on Motion : An Electronic Representation of the Manuscript Ms. Gal. 72, op. cit.,* folio page 113r.

9. A comparison of the folios 113 and 41/42 of *Galileo Galilei's Notes on Motion : An Electronic Representation of the Manuscript Ms. Gal. 72, op. cit.,* has shown that the compass marks of folio 41/42 match the ink spots on folio 113.

10. G. del Monte, *Meditantiunculae Guidi Ubaldi e marchionibus Montis Santae Mariae de rebus mathematicis,* Bibliothèque Nationale de Paris, Manuscript, Catalogue n° lat. 10246, Paris, *ca.* 1587-1592, folio page 236 ; a transcription of the text has been first published by G. Libri, *Histoire des Sciences Mathématiques en Italie,* vol. 4, Paris, Renouardi, 1838, 397ff.

11. A translation of this protocol is provided in P. Damerow, G. Freudenthal, P. McLaughlin, and J. Renn, *Exploring the Limits of Preclassical Mechanics,* New York, Springer, 1992, 337.

it really possible that Galileo discovered his error in the *De Motu* theory no earlier than 1607, as a belated consequence of the discovery of the law of fall which allegedly happened in 1604 ? To answer this question it is crucial to know when the Guidobaldo experiment took place, what Galileo's role was in this experiment, and how this experiment was related to Galileo's discovery of the law of fall. In view of the fact that the dating of Galileo's discoveries has always been considered as an important issue, it is surprising that even obvious clues pointing to earlier datings of the two events have been overlooked.

In 1602 Galileo wrote a letter to Guidobaldo del Monte claiming that he had proven the so called " Broken Chord Theorem "[12]. The related manuscripts[13] show that not only the final proof in the *Discorsi*[14] but also that all earlier attempts presuppose knowledge of the law of fall. Concerning the parabolic trajectory, Galileo himself states in a letter of 1632 to Cavalieri[15] that it was the result of a study of 40 years. The terminus ad quem for the discovery of the parabolic trajectory is therefore the year 1592, that is, Galileo claims that he studied the projectile trajectory already 10 years before he used the law of fall to prove the " Broken Chord Theorem " as it is documented by his letter to Guidobaldo. Is it credible that his remark in the letter to Cavalieri refers to the discovery of the parabolic trajectory or does it merely refer to the beginning of Galileo's study ? One indication that the first alternative might be true is provided by the use of a drawing of a parabola, similar to the drawing of a trajectory at the beginning of the 4th Day of the Discorsi, as an allegory for his discoveries in dedications to his friends such as the discovery of the Medicean planets[16]. The oldest use of this allegory that has been preserved is from 1599. The existence of this allegory makes it likely that Galileo discovered the parabolic form of the projectile trajectory not later than the law of fall, and even that something special might have happened in 1592.

Let us now directly look at the year 1592 to see if there is any indication for the discovery of the parabolic trajectory. It is well-known that in this year Galileo moved from Pisa to Padua where he began to work intensively on problems of motion. We know from letters between Galileo and Guidobaldo that the

---

12. *Cf.* " Letter of Galileo to G. del Monte, November 29, 1602 ", edited in G. Galilei, *Le opere di Galileo Galilei*, X, Florence, 1890-1909, 97-100, 20 vols.

13. *Cf.* e.g. folio page 186v of *Galileo Galilei's Notes on Motion : An Electronic Representation of the Manuscript Ms. Gal. 72, op. cit.* ; for the discussion of this folio page cf. W.L. Wisan, " The New Science of Motion : A Study of Galileo's *De motu locali* ", *Archive for the History of Exact Science*, vol. 13 (1974), 103-306, 177-179.

14. See G. Galilei, *Two New Sciences, op. cit.*, third day, theorem XXII (= prop. XXXVI), 211f.

15. " Letter of Bonaventura Cavalieri to Galileo, August 31, 1632 ", edited in G. Galilei, *Le opere di Galileo Galilei*, XIV, *op. cit.*, 378.

16. *Cf.* G. Galilei, *Le opere di Galileo Galilei*, XIX, *op. cit.*, 204.

latter invited Galileo to travel via Pesaro and pay him a visit[17]. There is no hint in the correspondence that Galileo did not follow this invitation.

It is furthermore well-known that from the very beginning of his presence in Padua Paolo Sarpi was one of his closest confidants, as far as matters of his research are concerned. Among the notes of Sarpi[18] around the year 1592 there are several entries concerning problems of motion, mechanics, and natural philosophy. In particular there is a note from the very year 1592 concerning the relation between the trajectory and the catenary and the reasons of the symmetry of these curves[19]. Although obviously independently written, this note follows so closely the protocol of Guidobaldo del Monte that it seems to be impossible that it was written without intimate information about the outcome of Guidobaldo's experiment. Thus, this experiment must have taken place before the entry in Sarpi's notebook. And if this should be true it can only have been Galileo travelling through Pesaro who had brought him a description of the outcome of the experiment and had convinced Sarpi of the plausibility of its theoretical implications which contradicted the commonly held views in the Aristotelian tradition.

However, if Galileo knew about the symmetric shape of the trajectory and its resemblance with the parabola as early as 1592, this would solve a much debated riddle : how could Galileo be so convinced of the law of fall although he was lacking a convincing proof ? Given Galileo's expertise, documented for instance by his early Theoremata of 1585-1587[20], he must immediately have recognized that the quadratic relation between distances and times is a direct implication of the theoretical reconstruction of the parabolic shape documented by Guidobaldo's protocol and Sarpi's note, that is, as a composition of forced horizontal and natural downward motion.

THE NATURE OF GALILEO'S DISCOVERY

We have assembled a number of hints pointing at an order and dating of Galileo's discoveries of the law of fall and of the parabolic trajectory different from the widely accepted ones. This immediately raises the question of why in spite of decades of Galileo studies these hints have been overlooked or have not been taken into account seriously. The answer is quite obvious. These hints fit neither into the picture of Galileo as a careful experimentalist performing

17. *Cf.* " Letter of G. del Monte to Galileo, February 21, 1592 ", edited in G. Galilei, *Le opere di Galileo Galilei*, X, *op. cit.*, 47.

18. P. Sarpi, *Pensieri naturali, metafisici e matematici*, Milano, Ricciardi, 1996.

19. *Cf.* J. Renn, P. Damerow and S. Rieger, " Hunting teh White Elephant : When and how did Galileo discover the law of fall ? ", in J. Renn (ed), *In Galileo in Context*, Cambridge, 2000, 13, 593-595.

20. *Cf.* G Galilei, " Theoremata circa centrum gravitatis solidorum ", in G. Galilei, *Le opere di Galileo Galilei*, I, *op. cit.*, 187-208.

precision measurements nor into that of a platonic theoretician applying rigorous mathematical deductions.

Galileo's arguments in favor of the common dynamical justification of his assumption of the parabolic shape of both the trajectory and the catenary, as we know them from Guidobaldo del Monte's protocol and Galileo's *Discorsi*, are weak and do not meet the theoretical requirements of classical physics. A simple experiment could have furthermore convinced him empirically that the catenary considerably differs from the parabola. Compared to Drake's claim[21] of having discovered the measurements of a crucial experiment with the inclined plane that led to the discovery of the law of free fall, the process of discovery as indicated by the evidence presented here does not correspond to the picture of a dramatic discovery. It rather seems that Galileo was convinced very early by weak arguments of the validity of the parabolic shape of the trajectory and the catenary as well as of the ensuing validity of the law of fall. It seems also, however, that he was for a considerable time unable to recognize the full theoretical implications of these insights.

This makes all identifications of a particular date as the date of the discovery questionable. What can it mean to date a discovery when the discovery is based on fallacious arguments and only in the course of a continuous process of theoretical work becomes a key theorem of a developed theory ?

Let us turn now to the question of Galileo's empirical research related to this hesitant discovery process which he only much later claimed so vividly as his most important breakthrough. Did he really never try to check the validity of his claim that the catenary and the projectile trajectory are both parabolic ?

Again, the codex Ms. Gal. 72 provides an answer. At some time Galileo carefully checked the relation between the catenary and the parabola, with the result that he discovered deviations and desperately tried to find an explanation. This is indicated by construction lines and numbers on folio page 107r which contains an empirically generated catenary and a carefully constructed parabola, together with attempts to find a rule for the differences between them[22]. The reverse page of the same folio, page 107v, contains again the figures of an empirical check of the catenary together with a drawing related to an attempted proof of the parabolic shape, to which we will return below. The close connection of both pages has been established by an ink analysis which shows that both pages have been written at the same time[23].

In the beginning we have mentioned the possible fallacies induced by the dominant concept of discovery in Galileo studies. It may be considered as

21. *Cf.* S. Drake, *Galileo : Pioneer Scientist, op. cit.*

22. These construction lines are published and related to the figures in the border for the first time in J. Renn, P. Damerow and S. Rieger, " Hunting teh White Elephant : When and how did Galileo discover the law of fall ? ", *op. cit.*

characteristic of this pattern of research that it is this very page 107v which Drake and others following him considered as documenting the crucial experiment for the discovery of the law of free fall by use of an inclined plane, misinterpreting the precise geometrical construction on this page as the raw sketch of a water container of a time measuring device[24].

Let us finally turn to the question of why Galileo stuck to his fallacious conviction of the equal shape of the trajectory and the catenary, in spite of empirical evidence to the contrary. The answer again follows quite clearly both from Galileo's manuscripts and the published *Discorsi*. In the sequel of his work he supplemented his original intuitive interpretation of the shape of the catenary and the projectile trajectory as the effect of similarly acting forces by at least two promising attempts to prove the parabolic shape of the catenary.

The first attempt tries to approximate a hanging chain by a string with a number of weights attached in the hope of showing that these weights always take a position on a parabola. Evidence is provided by the folio pages 132r and v. On these pages Galileo tries to show that a string with three hanging weights fits to a parabola. These pages however leave open the question of how Galileo could know which shape a string with three hanging weights assumes. It was a sensation for us when we discovered by analyzing the construction lines on folio page 107v that the alleged container for time measurements by means of a water clock was actually an ingenious construction of the lowest position of the center of gravity of a hanging string with three attached weights[25]. This attempt led him to an argument proving that a hanging chain can never be completely rectilinear, however great the applied forces are, an argument which he included at the end of his treatment of projectile motion in the *Discorsi*. He gave up, however, his attempt to transform this approach into a rigid proof of the parabolic shape because he failed to complete it.

The situation is different for the second attempt, also documented by a page of the codex Ms. Gal. 72, folio page 43r, on which Galileo sketched a proof of the parabolic shape of the catenary completely different from his earlier attempt. This proof is based on Galileo's theory of the strength of materials and hence links the two new sciences presented in the *Discorsi*. It turned out to be even more powerful because it seemed to show that the same argument does not only comprise the parabolic shape of the trajectory and the catenary but also the parabolic shape of a beam of equal stability.

23. For the results of the ink analysis *cf.* the report *Pilot Study for a Systematic PIXE Analysis of the Ink Types in Galileo's Ms. 72*, Berlin, Max Planck Institute for the History of Science, 1996 (Preprint n° 54), and for the interpretation *cf.* J. Renn, P. Damerow and S. Rieger, " Hunting teh White Elephant : When and how did Galileo discover the law of fall ? ", *op. cit.*

24. *Cf.* S. Drake, *Galileo : Pioneer Scientist*, *op. cit.*, 9-12.

25. *Cf.* J. Renn, P. Damerow and S. Rieger, " Hunting teh White Elephant : When and how did Galileo discover the law of fall ? ", *op. cit.*, 74ff.

There is overwhelming evidence, widely neglected by historians of science, that Galileo was convinced by the validity of this proof until the end of his life and that only his death prevented him from making this argument the focus of the unwritten fifth day of the *Discorsi*. We know for instance from a letter of Galileo's publisher Elzevier that Galileo had already announced this intended completion of the *Discorsi*[26]. We also have a report by Viviani about the essence of Galileo's proof as sketched on folio page 43r of Ms. Gal. 72 and about Galileo's intention to include this proof in the *Discorsi*[27].

We have hence to admit that in a situation where empirical evidence contradicted rational arguments Galileo — at least in this case — decided to believe more in the argument than in the empirical evidence. And he had good reasons for that. In a letter to Guidobaldo he argued : " (…) that when we begin to deal with matter, because of its contingency the propositions abstractly considered by the geometrician begin to change (…) "[28].

Summing up :

- Galileo found the law of fall by a qualitative experiment on projectile motion.

- His discovery of the law of fall and of the parabolic trajectory happened much earlier than commonly assumed but they did not have the path-breaking and dramatic consequences that are usually associated with them.

- He did not reject the *De Motu* theory, as it is usually assumed, because it was incompatible with the results of systematic experiments on motion which he carried out.

- He performed careful experiments with chains ; in particular, the alleged crucial experiment by which he supposedly found or confirmed the law of fall was in fact an experimental control of the parabolic shape of the catenary.

- In spite of empirical evidence to the contrary he stuck to his identification of catenary and trajectory.

- Galileo searched and found a common dynamical understanding of three issues that seemed and seem unrelated also today from the viewpoint of classical physics : the parabolic shape of the trajectory, the parabolic shape of the beam of equal stability, the parabolic shape of the catenary.

- He planned to complete the *Discorsi* in a way which does not fit the perception of this work as founding classical mechanics.

26. " Letter of L. Elzevier to Galileo, November 1, 1637 ", edited in G. Galilei, *Le opere di Galileo Galilei, op. cit.*, XVII, 211.

27. *Cf.* V. Viviani, *Quinto libro degli Elementi di Euclide, ovvero scienza universale delle proporzioni, spiegata colla dottrina del Galileo*, Florence, Condotta, 1674, 105f.

28. " Letter of Galileo to G. del Monte, November 29, 1602 ", *op. cit.*

These findings challenge the notion of discovery widely held by historians of science. Galileo had a strong, well thought-out understanding of his own discoveries, but this understanding differs considerably from the understanding of his achievements held by many historians of science. Galileo performed careful experiments but neither were they those ascribed to him, nor did he draw the conclusions from them that he should have drawn according to the assumption that they would explain his " discoveries ". Galileo was convinced that he had strong theoretical justifications for his most prominent discoveries but neither do they justify what they should do according to the standard interpretation of his discoveries, nor do they fit an understanding of " justification " that corresponds to the modern distinction of the context of justification and the context of discovery.

Notes pour une traduction intégrale de l'essai contenu dans les *De motu antiquiora* de Galilée

À la mémoire de Israël Edward Drabkin

Contrairement aux apparences, l'essai en 23 chapitres, que la traduction anglaise partielle de Drabkin a bien fait connaître, est un ouvrage, en deux livres, bel et bien achevé. Testant dans son entourage ses idées, ses observations ainsi que ses expériences, notre jeune titulaire de mathématique du *Studio di Pisa* réalise les difficultés de se faire comprendre. Se relisant attentivement, l'ardent disciple d'Archimède réalise aussi que son exposé théorique, au premier livre, manque de clarté et de rigueur. Or, après en avoir achevé entièrement une seconde version comportant cette fois 16 chapitres, il va abandonner son projet de publier.

Raymond Fredette

Je m'adresse ici plus particulièrement à celles et ceux parmi vous qui, pour vous familiariser avec la dynamique pisane de Galilée, avez utilisé la traduction que I.E. Drabkin[1] faisait paraître en 1960. Le menu principal de cette traduction[2] est un essai constitué par les 23 chapitres que Favaro[3] a placés en tête de l'édition critique (EN I, 251-340), 89 pages de texte représentant un peu plus de 50% des matières de l'autographe. Le manuscrit nous restitue aussi des contenus analogues à ceux de l'essai sous formes dialoguées (EN I, 367-408), représentant 41 pages, soit un quart des matières. Drabkin avait clairement établi à l'époque que ce dialogue était resté inachevé et que sa rédaction était antérieure à celle de l'essai[4]. Par contre, le restant des matières de l'autographe, soit 25 pages de texte, auxquelles il faut ajouter leur dix pages de notes (EN I, 341-366 + 409-419), représente rien de moins que 20% des matières de l'autographe. Or ce matériel ne fait l'objet par Drabkin que de quelques remarques à l'effet qu'il s'agit de *reworkings* de l'essai principal dont il donne la traduction. Ma communication a pour objet de faire valoir la nécessité d'un

---

1. *Cf.* son " Éloge " par Ed. Rosen, *Isis*, 56 (1965), 434-437.
2. G. Galilei, *On Motion and On Mechanics*, Madison, The University of Wisconsin Press, 1960, 1-131.
3. G. Galilei, *Opere*, I, a cura di A. Favaro, G. Barbera, Firenze, Edizione Nazionale, 1890.
4. " A Note on Galileo's *De motu* ", *Isis*, 51 (1960), 271-277.

examen de ce matériel dans son intégralité, si on veut apprécier à sa juste valeur la dynamique pisane.

Lorsqu'on prend connaissance de l'essai dans sa version originale, on se trouve devant un travail, pour ainsi dire achevé et complet, en 23 chapitres. On peut suivre dans les moindres détails notre jeune Galilée relisant son essai. Il insère des notes marginales, des additions, de courtes réflexions en passant, des références précises à des textes d'Aristote qu'il n'avait pas encore exploités. Il corrige son propos entre les lignes, il biffe des passages en les remplaçant par d'autres qu'il inscrit quelque fois sur des feuillets à part, etc. Et on constate qu'une fois son travail de relecture terminé, Galilée s'est vu contraint d'entreprendre de réviser de façon majeure son essai.

Il y a lieu ici de rappeler un fait étonnant. Dans la traduction, au ch. 19, p. 88, de cette version I, on lit : *For we know definitely, from what was proved at the beginning of this book, that speed and slowness are a consequence of weight and lightness.* Galilée a écrit : *cum certo sciamus, ex demonstratis in primo libro, velocitatem et tarditatem, gravitatem et levitatem sequi.* (EN I, 318.9-10). Comment Drabkin a-t-il pu décider de traduire *in primo libro* comme s'il avait lu *in principio libri* reste à ce jour pour moi un grand mystère. Or cette initiative a contribué à occulter un fait capital, à savoir, que l'essai comporte bel et bien dans l'esprit même de son auteur une structure en deux livres. Un premier, d'allure essentiellement théorique, comportant les 13 premiers chapitres de la traduction de Drabkin, suivi d'un second, consacré à des questions adjacentes particulières, allant du fameux chapitre 14 sur le plan incliné jusqu'à un 23ᵉ, sur la balistique. Le passage cité révèle par ailleurs deux choses. Première-ment, constatez qu'il s'agit d'un lieu dans le texte où l'auteur nous renvoie à un autre passage de son texte, ici, le chapitre 7, intitulé : la cause de la vitesse et de la lenteur du mouvement naturel. Or, les 89 pages qui composent cette version I de l'essai nous prodiguent rien de moins que 60 de ces renvois, celui-ci, au chapitre 19, étant le 49ᵉ. Et, lorsqu'on soumet à une opération de coor-dination l'ensemble constitué par toutes et chacune des unités d'information concernées, *i.e.* les 60 renvois avec leurs répondants, le résultat obtenu révèle que la v. I de l'essai est définie par un ensemble d'exactement vingt trois cha-pitres. Et l'ordre dans lequel chacun exige d'être lu est strict, total et sans écart, cet ordre est unique, il tient compte de tous les chapitres et n'en admet aucun autre. Nous voilà donc devant un essai rigoureusement construit, ordonné, achevé et matériellement complet dans sa rédaction d'ensemble. En deuxième lieu, on peut constater que Galilée utilise encore ici ce double discours aristo-télicien de la pesanteur et de la légèreté, de la vitesse et de la lenteur. Or, les révisions ont pour objet principal d'éliminer ce double discours .

Un premier travail de révision va vite avorter (EN I, 341-343). " ...si parfois je m'exprimais selon l'usage communément répandu (la guerre des mots a en effet peu d'intérêt pour notre propos) et que je parlais du pesant et du léger, et de haut et de bas, on comprendra par cela moins pesant et plus pesant, et plus près du centre et plus éloigné ; jusqu'à ce qu'il soit possible, quand viendra l'occasion, d'être plus nuancé à ce sujet ". (EN I, 342.3-11).

Cette occasion, elle ne viendra pas. Galilée, à Pise, — nous sommes en 1590, il n'a que 26 ans —, va inaugurer une activité à laquelle, en définitive, il sera contraint de consacrer, pendant encore cinquante ans, une partie considérable de son énergie : faire valoir qu'on doit penser autrement que nous l'a appris Aristote.

Revenons au texte. Nous disposons pour ce faire de son autographe. Relisant par dessus son épaule, on constate qu'il a relu son essai avec grande attention. Sur 23 chapitres, 16 ont été retouchés ici et là. Seuls les ch. 4, 5, 9, 13, 16, 18 et 23 n'ont pas eu besoin de l'être et ce sont tous des chapitres secondaires. Il a élaboré toute une théorie de la vitesse uniforme de translation des corps qui repose sur l'assertion que ces corps sont mus conformément à ce que nous, nous appellerions leurs " pesanteurs spécifiques ", relatives aux différents milieux dans lesquels leurs mouvements peuvent avoir lieu. Dans un milieu dont " l'épaisseur " —*crassities*, nous dirions bien entendu la densité — serait égale à la pesanteur par unité de volume du corps, ce dernier n'aurait évidemment ni vitesse ni lenteur. Il se tiendrait en équilibre et au repos dans ce milieu. Cependant, si " l'épaisseur " du milieu était inférieure, l'équilibre serait rompu. Le corps de par " l'excès de sa pesanteur " serait mu " naturellement et vers le bas " à une vitesse ou une lenteur uniforme.

Par contre, si " l'épaisseur " du milieu était supérieure à la pesanteur par unité de volume du corps, encore là l'équilibre serait rompu ; mais, cette fois, par un " excès de sa légèreté ", il serait mu " violemment et vers le haut ", toujours cependant à une vitesse ou une lenteur uniforme. On voit bien ici comment le langage de Galilée conserve aux " vieux mots " le droit et le pouvoir d'exprimer des " choses nouvelles ". Il n'y a plus, à proprement parler, de mouvement " naturel " vers le haut. La propriété pour le mouvement d'être naturel ne s'applique que sur le fait que tous les corps, étant matériels, sont donc tous plus ou moins pesants. Tout mouvement vers le haut est violent. Dans le vide, tous les corps devraient être mus " naturellement " vers le bas à des vitesses différentes spécifiquement mais uniformes, en vertu, cette fois, de ce que Galilée appelle leur " pesanteur propre, essentielle et absolue " (EN I, v. I, ch. 10, 281.13-17).

Comme on peut le constater, des écarts fondamentaux d'avec la théorie d'Aristote sont exprimés avec beaucoup de clarté dans le langage même de la pensée qu'ils dénoncent. Souvenez-vous comment les démonstrations du chapitre 8 sont doubles, une pour le mouvement vers le bas, une autre pour le mouvement vers le haut.

Imaginez-vous un instant, quel effet de telles façons de dire ont pu avoir dans les oreilles d'un Francesco Buonamici ? Au moment où Galilée rédige sa dynamique pisane, Buonamici publie son monumental *De Motu*. Et il n'est pas impossible que l'ancien élève devenu collègue ait contribué par ses idées incongrues à convaincre[5] le maître de la nécessité de rédiger son oeuvre synthèse. Galilée doit montrer où Aristote se trompe. Il n'y a là rien d'évident. Il passe à l'attaque. Ce sera ce que j'appelle la version III. Et après plus de 11 pages de textes révisés[6], voici en bonne et due forme sa déclaration officielle de guerre. Treize lignes d'une grande virulence et qui vont en fin de compte marquer l'histoire : " Jusqu'ici nous n'avons même pas parlé du léger, mais seulement du pesant et du moins pesant ; c'est pourquoi ce lieu fournit l'occasion d'examiner si c'est à bon droit ou bien à tort que nous avons fait cela. C'est ainsi que si Aristote et le reste des philosophes s'étaient contentés d'entendre par léger ce que nous nous appelons moins pesant, il ne nous pèserait pas d'admettre, nous aussi, cette appellation de léger : au contraire, ils ont voulu (ne se contentant pas de comprendre par léger ce qui est moins pesant) que soit donné, en plus, un certain corps léger, qui serait simplement tel et serait privé de toute pesanteur. Fuyant cela avec plus d'horreur qu'un chien un serpent, nous sommes contraints d'expulser tout à fait complètement jusqu'au léger lui-même. C'est pourquoi suivant en cela l'opinion des anciens qu'Aristote essaie vainement de renverser au livre IV du *De Caelo*, nous examinerons, non seulement les réfutations d'Aristote qu'il y a là, mais encore ses preuves, en prouvant d'une part ce qui a été réfuté, en réfutant d'autre part ce qui a été prouvé. Et nous nous acquitterons de cela lorsque nous aurons exposé l'opinion d'Aristote "[7].

Et le résultat de ce second travail de révision sera mené à terme avec un soin extrême. Une analyse fine de cette révision révèle[8] une réorganisation de tout l'ordre des matières de son premier livre, dans un exposé qui en accroît considérablement la rigueur au plan formel et qui élimine toute équivoque sur le sens résolument original et la portée radicale de sa première entreprise théorique en dynamique.

Que voit-on ? D'une part, que les chapitres 1 à 6 ainsi que 12 de la version I qui faisait l'objet de la traduction de Drabkin, sont éliminés et nommément remplacés par les dix sections que Favaro a publiées dans l'ordre où elles se trouvent ensemble dans l'autographe. Sachez que nous faisons là l'acquisition

---

5. Dans sa dédicace F. Buonamici écrit : *Occasio vero scribendi voluminis ab ea controversia sumpta est, quae in Academia Pisana inter nostros collegarumque auditores exorta est de motu elementorum.* Cf. M.-O. Helbing, *La Filosofia di Francesco Buonamici professore di Galileo a Pisa*, Pisa, Nistri-Lischi, 1989, 375.

6. *Cf.* EN I, 344-355 *i.e.* les 7 premières sections de la v. III.

7. *Cf.* v. III, #8, EN I, 355.7-20. C'est dès le départ que Galilée prend le parti des atomistes.

8. *Cf.* R. Fredette, *Les De motu " plus anciens " de Galileo Galilei : prolégomènes.* Thèse présentée en vue de l'obtention du grade de Philosophiae Doctor, Montréal, Institut d'Études médiévales de l'Université de Montréal, 1969, 237-282. Cette thèse est accessible en format ISO tant pour Macintosh que pour Windows à la rubrique Datenbanken du site du Max-Planck Institut für Wissenschaftsgeschichte : http : // www2.mpiwg-berlin.mpg.de/.

de 4 chapitres[9] entièrement nouveaux, *i.e.* qui n'ont aucun homologue dans cette première version de l'essai. D'autre part, on peut voir que les chapitres 7, 8, 9, 10, 11 et 13 de la v. I, eux, ont été retenus dans cet ordre et intégralement par Galilée révisant de fond en comble l'exposé théorique du premier livre de son essai. En effet, on constate que les renvois qu'on trouve dans sa révision en 10 sections, se réfèrent à ces chapitres, dont on sait qu'ils sont déjà rédigés, en utilisant le mode indicatif au futur. C'est donc dire que, après les avoir relus et corrigés, notre jeune physicien aura jugé bon de retenir ces chapitres. Ainsi, chacun des 13 chapitres de la v. I, qui constituaient en un livre complet et achevé la version originale de son exposé théorique, a été affecté par le travail de révision de la v. III. Résultat : Galilée a entièrement renouvelé le premier livre de son essai. Il est à nouveau achevé et complet. Et il comprend maintenant seize chapitres[10].

La v. III fait état de tous les chapitres théoriques de la v. I du premier livre, sauf un. Il s'agit du fameux chapitre 6. Notre disciple d'Archimède et auteur de *La Bilancetta*, si fier qu'il était de penser pouvoir réduire les mobiles naturels, tant *sursum* que *deorsum*, aux poids de la balance, se voit contraint, en toute rigueur, de relativiser l'importance de cette idée maîtresse du projet initial, pilier central de son livre I original que la traduction de Drabkin a bien fait connaître[11]. Deuxièmement, voici en quels termes Galilée annonce le tout nouveau chapitre 10, EN I, 362.30-363.4, qui vient clore ses révisions : " Or, je croirais que l'erreur de ceux qui ont pensé que le mouvement des corps qui s'éloignent du centre est naturel, prend naissance dans le fait qu'ils n'ont pas pu trouver la cause externe par laquelle ces mobiles sont mus ; et que, pour cela, ils ont été forcés d'en supposer une interne, qu'ils ont appelée légèreté. C'est pourquoi, afin que nous détruisions une erreur de ce genre, nous allons tout de suite nous dépêcher d'expliquer comment des choses qui sont mues vers le haut sont mues par une cause extrinsèque, à savoir par expulsion de la part du milieu lui-même ".

Qu'est-ce à dire ?

9. *Cf.* EN I, d'abord la section #4, 348.24-350.3, un *lemma ad sequentia* qui vient réparer une grave faute de rigueur au ch. 8 (*cf.* 264.11-13, Drabkin, n. 6, 28 ; R. Fredette, " Galileo's *De Motu Antiquiora* ", *Physis*, XIV, fasc. 4 (1972), 344) ; puis, #7, 352.31-355.5, il n'existe aucun mouvement naturel vers le haut ; et #9, 361.6-363.4, le mouvement vers le haut de la part du mobile ne peut pas être naturel ; et enfin, #10, 363.5-366.5, les choses qui jusqu'à présent ont été dites être mues vers le haut naturellement ne sont pas mues par une cause interne mais externe, à savoir par le milieu lui-même, par expulsion.

10. Ce sont, dans l'ordre, les dix chapitres de la révision proprement dites, EN I, 344-366, suivis des six chapitres dûment corrigés de la version originale que l'auteur conserve, soit EN I, 260.3-289.24 et 294.14-296.4.

11. Depuis la présente prestation à Liège, une double traduction intégrale française-anglaise de l'autographe des *De motu antiquiora* est en marche. Pour l'anglaise, je bénéficie du support du Max-Planck-Institut für Wissenschaftsgeschichte de Berlin, aux fins d'une version électronique qu'on trouvera bientôt sur leur site internet. Est également en chantier ma traduction d'une nouvelle édition des textes, produite par E. Giusti, et qu'A. Segonds, le directeur général des Belles-Lettres, m'a invité à lui soumettre.

Lui aussi, il a désigné du nom de légèreté cette *virtus impressa*, cette force impresse. Vous vous souvenez comment il y fait appel, d'abord au ch. 17, comme cause interne du mouvement des projectiles, et ensuite au ch. 19, où il s'en sert pour " sauver sa théorie " de la contradiction avec les phénomènes d'accélération, phénomènes indéniables et qu'il a confirmés par expérimentation[12], mais qu'il interprète comme de la vitesse progressivement déretardée. Il ne peut manquer de voir ces confusions. Décidément, son projet n'est pas encore prêt pour la publication. Il remet ça et retourne au travail. Il est bien armé. Il dispose déjà de l'arsenal d'outils qui le rendront célèbre, à savoir sa connaissance théorique et pratique d'Euclide et Archimède et son habilité révolutionnaire à explorer l'étude de la nature du mouvement de façon expérimentale, à l'aide d'objets techniques, le plan incliné, bien sûr, mais aussi le pendule[13].

Le long de cette trajectoire qui mène la pensée occidentale d'Aristote à Newton, Galilée tient une place à la fois capitale et provisoire, un lieu nécessaire mais néanmoins éphémère[14]. Lorsque la science moderne prétend qu'elle a un Galilée pour père, elle s'imagine le plus souvent qu'elle tient son savoir de lui ; elle se trompe. Pas une seule des lois de la science galiléenne ne sera retenue sous la forme ou la signification que Galilée lui donnait. Ce que la science moderne tient de Galilée n'est pas son savoir mais son origine, c'est-à-dire l'autorisation de parler et de procéder autrement. Or, ce n'est pas moins que sa carrière toute entière que Galilée devra investir pour obtenir le privilège d'octroyer cette autorisation, c'est-à-dire de rendre complète la rupture avec le vénérable héritage de la pensée d'Aristote.

12. *Cf.* T.B. Settle, " Galileo's Use of Experiment as a Tool of Investigation ", dans E. McMullin (ed.), *Galileo : Man of Science*, New-York, Basic Books, 1967, 315-337.

13. *Cf.* au ch. 22, EN I, 335.22-26 et le mémorandum qui lui correspond, en 413.1-3.

14. Cette idée du caractère éphémère et provisoire de la science galiléenne, une idée enfin aujourd'hui acceptable et acceptée, c'est à T.B. Settle que nous la devons. *Cf.* T.B. Settle, *Galilean Science Essays in the Mechanics and Dynamics of the DISCORSI* . A Thesis Presented to the Faculty of the Graduate School of Cornell University for the Degree of Doctor of Philosophy, June 1966. Cette thèse, qui comme la mienne n'a jamais été publiée, est maintenant accessible en format ISO tant pour Macintosh que pour Windows à la rubrique Datenbanken du site du Max-Planck Institut für Wissenschaftsgeschichte : http://www.mpiwg-berlin.mpg.de/. Voir notamment ses conclusions, 248-255. Il n'est peut-être pas inutile de rappeler ici que le titre de l'article inaugural de la carrière de Settle, est *An Experiment in the History of Science*. Il a été publié — cela n'est pas innocent — non pas dans un périodique d'historien, mais dans *Science*, 133, #3445 (1961), 19-23. T.B. Settle, reproduisant avec d'excellent résultats telle que décrite dans les *Discorsi* l'expérience du plan incliné, prouve expérimentalement qu'A. Koyré, alors menant une carrière triomphante aux USA, avait tort de prétendre, sans se donner la peine de le tester lui-même, qu'il suffisait de lire le texte de Galilée pour savoir que cet expérience n'était qu'une expérience de pensée. *Cf.* P. Bozzi, C. Maccagni, L. Olivieri, T.B. Settle, *Galileo e la scienza sperimentale*, a cura di M. Baldo Ceolin, Padova, Dipartimento di Fisica " Galileo Galilei ", 1995, où T.B. Settle fait état des expérimentations classiques sur le plan incliné, sur les pendules tant harmoniques qu'interrompus, sur la percussion, mais aussi celles moins connues des *bicchieri canterini*, sur les phénomènes oscillatoires de la résonance. Il serait grandement temps que nous ayons entre historiens des sciences des rencontres dans le but explicite de valoriser plus que nous n'avons su le faire jusqu'ici les sensibilités épistémologiques des *scienziati* parmi nous, par comparaison avec celles de la majorité d'entre nous qui sommes plutôt des *letterati*.

THE CONTRIBUTION OF MATHEMATICAL, PHILOSOPHICAL AND TECHNICAL CULTURES TO 16TH CENTURY HYDRAULICS

Alessandra FIOCCA

The Late Renaissance saw a profound change in the overlapping of competencies among various professional categories : philosophers, engineers and architects, mathematics and astronomy teachers at the Academies, Universities and Religious Colleges. There are numerous examples of mathematics teachers and philosophers engaged in technical fields and, vice-versa, technicians with a scientific and mathematical background deeply rooted in classical antiquity.

The circulation of ideas and the exchange of knowledge among various members of any society is considered essential to the development of new approaches to the study of the physical world. An exemplary case of such transmission is found in the management of the small torrential Reno River in Italy during the first half of the 16th century and on into the following centuries[1].

The present article introduces the problem, provides the intellectual profiles of some of those involved and the trends in their thought. The aim is to identify the changes in mentality and original insights which arose. We will see that the distance between Renaissance hydraulic tradition and hydraulic science — the birth of which is historiographically considered coincident with the publication of *Della misura delle acque correnti* (1628) by Benedetto Castelli — is not so great after all[2].

---

1. On the tormented story of the River Reno see the reconstruction by A. Giacomelli, " Appunti per una rilettura storico-politica delle vicende idrauliche del Primaro e del Reno e delle bonifiche nell'età del governo pontificio ", *La pianura e le acque tra Bologna e Ferrara (un problema secolare)*, Cento, 1983, 101-254.

2. There are several titles on the history of hydraulics ; among these I found the following very interesting : C.S. Maffioli, *Out of Galileo. The Science of Waters 1628-1718*, Rotterdam, 1994 ; *Hydraulics and Hydraulic Research. A Historical Review*, in G. Garbrecht (ed.), Rotterdam, 1987 ; H. Rouse and S. Ince, *History of Hydraulics*, Iowa Institute of Hydraulic Research, 1957 ; A.K. Biswas, *History of Hydrology*, Amsterdam, London, 1970.

The hydrography of the lower Po River Valley prior to the 16th century was quite different from what it is today. Originally, flowing down from the Lombard Plain, the Po River ran south of Ferrara and, beyond the city, divided into two branches : the " Po di Volano " to the north-east and the " Po di Primaro " to the south-east.

The city of Ferrara was the hub of fluvial traffic from the Adriatic Sea to the Lombard Plain. Indeed, the river was not only associated with trade prosperity but also with the city's military safety.

In the middle of the 12th century the Po breached its banks 15 miles above Ferrara, creating a new branch north of the older, main branch. This new branch, today the actual course of the river, was called the Po di Venezia. Despite the division of the waters, the old branch with its two secondary branches, Volano and Primaro, remained deep and navigable for several centuries.

Seven streams running down from the Apennines terminated in the broad Padusa marsh extending south of the Primaro branch all the way to the Adriatic Sea. In time the terrigenous sediments from these torrential rivers partially reclaimed the marsh. However, at different times, people let the torrents flow into the various branches of the Po in an attempt to speed up the natural course : the Santerno in 1460, the Lamone in 1504, the Senio in 1537 into the Po di Primaro ; the Reno in 1526 into the Po di Ferrara. The latter produced the most abrupt, most disastrous effects.

Inlet of the Reno into the Po di Ferrara was arranged between Alfonso I d'Este, Duke of Ferrara, and the city of Bologna, the agreement dated the 5th of December, 1522. Canalization of the Reno quickly transformed the lands affected by its expansion and strongly reduced the level and extension of the marshes. Nevertheless, the first disappointments were not long in coming : the repeated breaching of its banks proved how inadequate the new banks were.

However, the most serious damage was the adverse effect this inlet had on the entire hydrographical system of the Dukedom of Ferrara : the strong impetus of the water flowing in from the Reno cut across the current of the Po ; as these waters expanded into the wider river bed, they slowed down the current depositing terrigenous sediments and elevating the level.

To prevent total loss of navigation, land drainage and military defense altogether Dukes Ercole II and his son Alfonso II repeatedly attempted to negotiate a new agreement with Bologna regarding the Reno. On-site technical visits with the supervisor of the Papal delegates and the participation of technicians from both parties began and continued throughout the 16th century and for the centuries to come.

It is possible to follow the negotiations with Bologna over the Reno River, the reasons why everything fell through in the historical reconstruction by the philosopher Francesco Patrizi of Cherso who wrote it all down in the 1580s[3].

Francesco Patrizi is quite an interesting Platonic philosopher. He is well known for his violent anti-peripatetic debate found in his four-volume work *Discussiones peripateticae*[4]. He is, however, less known as a land-reclamation man and hydraulic engineering planner[5]. His biography is mainly derived from the letter he wrote to his friend Baccio Valori on January 12, 1587[6].

Patrizi was born on April 25, 1529 in the Dalmatian Isle of Cherso, then under the jurisdiction of the Republic of Venice. In 1547 he attended medical school at the University of Padua but he soon started to study philosophy on his own. In 1560 he was assigned to read the Ethics of Aristotle to Giorgio II Contarini, Count of Zaffo who, a year later, sent Patrizi to Cyprus to govern his lands. Patrizi remained in Cyprus for 6 years. He had administrative tasks but he also found time to cultivate his cultural interests, collecting a rich library of ancient Greek codices on philosophy, theology and science. Today a part of this library is housed in the Escurial Library[7]. He also involved himself in the reclamation of the lands surrounding the village of Calopsida, near Famagosta, digging channels and building banks so that cotton could be grown.

In 1568 he returned to Venice and resumed his studies. A Spanish gentleman, Diego Hurtado de Mendoza y de la Cerda, Viceroy of Catalonia, asked Patrizi to become his personal " philosopher " and accompany him to Barcelona. Don Diego, who translated the pseudo-Aristotelian mechanical questions into Spanish and was one of the interlocutors for Nicolò Tartaglia in his work *Quesiti et inventioni diverse*[8], was also one of the men Filippo II contacted to establish his Library. Patrizi only remained in Spain for 6 months, occupying himself, among other things, in the trade of codices and manuscripts.

From 1571 to 1573 he was once more in Venice where he started out on an entirely new endeavor as editor and book-seller with his nephew Giovanni Franco. Among the works he published under the name " All'Elefanta ", wor-

3. See the letter by Patrizi to the Duke Alfonso II, on July 5, 1580, in *Archivio di Stato di Modena* (hereafter called ASM), Confini dello Stato, 34A. This letter is now available in print ; see A. Fiocca, " Francesco Patrizi e la questione del Reno nella seconda metà del Cinquecento ", in P. Castelli (ed), *Francesco Patrizi filosofo Platonico nel crepuscolo del Rinascimento*, Firenze, 2001, 245-277.

4. F. Patrizi, *Discussiones peripateticae*, vol. 4, Basel, 1581.

5. A recent book on F. Patrizi is : C. Vasoli, *Francesco Patrizi da Cherso*, Roma, 1989.

6. This letter is published in : F. Patrizi da Cherso, *Lettere e opuscoli inediti*, in D. Aguzzi Barbagli (ed.), Firenze, 1975, 45-51.

7. M. Muccilli, " La biblioteca greca di F. Patrizi ", *Bibliothecae Selectae da Cusano a Leopardi*, Firenze, 1993, 73-118.

8. N. Tartaglia, *Quesiti et inventioni diverse*, Venice, 1546.

thy of mention is *Della proportione et proportionalità communi passioni del quanto* by Silvio Belli, concerning Euclidean theory of proportions[9].

After his activity as publisher and book-seller had come to naught, Patrizi decided to return to Spain in an attempt to recover the codices previously left behind by some servants. He also wanted to propose his own plan for the rearmament of 600 galleys to Filippo II who was, at that time, waging war against the Turks. He felt he was the right person to carry out such a plan as, during his long stay in Cyprus, he had acquired an in-depth knowledge of Turkish war tactics. Patrizi most likely gleaned his expertise in warfare from his association with the jurist and bibliophile, Girolamo Maggi, military engineer in Cyprus while Patrizi was there[10]. The galley rearmament project came to naught but Patrizi was able to finalize sale of his Greek manuscripts to Filippo II. The catalogue of volumes the Spanish King purchased from Patrizi reads like a patrimony of encyclopaedic erudition and Patrizi's own Cyprian library distinguishes itself for its rarity and uniqueness as well as for its vastness in terms of time and topics. Patrizi's project to bring to light the encyclopaedia of Platonic sciences, at the core of which lay music and mathematics, is also clearly reflected in the composition of this library[11].

From 1577 to 1592 Patrizi was in Ferrara, at the Court of the Estensi Dukes, with an appointment to teach Platonic philosophy at the local University. This was his most prolific period intellectually and he developed and completed his project to renew philosophy by rediscovering Platonic, neo-Platonic and Hermetic traditions.

During the course of his life Patrizi published 35 works (19 originals and 16 translations or works which he annotated), 5 works were published after his death, as of 1970 11 were still unpublished and 7 have been lost. In 1587 he published *Della nova geometria*, an attempt to improve on the work by Euclid and Proclo by elaborating a new, more satisfactory geometrical method in keeping with his conception of science. In 1591 he published *Nova de universis philosophia* where he accepted the theory of earth rotation and disclaimed the possibility of any motion by the fixed star heaven[12]. In 1592 he went to the Papal Court in Rome to teach Platonic Philosophy at the University. Five years later he died in Rome and was buried in the Church of Saint Onofrio, near the grave of Torquato Tasso who was, incidentally, his antagonist in an argument on heroic poetry.

9. C. Marciani, " Un filosofo del Rinascimento editore-libraio : F. Patrizio e l'incisore G. Franco di Cherso ", *La Bibliofilia*, 72, n° 2 (1970), 177-198, n° 3, 303-313.

10. Maggi was also co-author with F.G. Castriotto of the three-volume *Delle fortificazioni delle città*, Venice, 1564, considered the fullest treatise on fortifications published to that time.

11. For the catalogue of the Greek manuscripts which Filippo II bought from Patrizi, see M. Muccilli, " La biblioteca greca di F. Patrizi ", *op. cit.* , 83-91.

12. C. Vasoli, " Gli astri e la corte (l'astrologia a Ferrara nell'età ariostesca) ", *La cultura delle corti*, Firenze, 1980, 129-158.

In 1975 Danilo Aguzzi Barbagli published some of Patrizi's as yet unpublished works, bringing to light his technical profile. These works include : *Frammento sul Panaro e sul Po* ; *Discorso sopra lo stato del Po di Ferrara* (1579) ; *Risposta alla scrittura di D. Scipio di Castro sopra l'arrenamento del Po di Ferrara*, (c. 1579) ; *Relazione di quanto ha negotiato a Ravenna per Sua Altezza con Monsignor Reverendissimo San Felice* (1580). To this list we must add theree manuscripts by Patrizi regarding the hydraulic problems of Ferrara : a report made to Alfonso II on July 5, 1580 regarding the " old negotiation concerning the River Reno and the River Po from the beginning until 1556 " and two letters written to the same Duke on November 28, 1580 and December 9, 1581 : in these letters Patrizi explains, respectively, the theoretical basis for his proposals and the details of his proposal to separate the waters of the Reno from those of the Po[13].

I do not want to explain the technical details of Patrizi's proposals to resolve the Reno problem at this time. Instead I would like to highlight some insights into experimental methodology found in his writings[14]. For example, Patrizi confutes the opinion of Papal technician Scipio di Castro indicating that all substances carried by the river, whether heavy or light, lie on the surface of the water[15]. According to Patrizi, the theoretical basis to disprove this opinion is that nature does not allow a single cause to produce the same effect on contrary objects. However, Patrizi does admit that this argument is highly subtle and thus suggests performing three experiments. Indeed, he says that " experience is the best and strongest of reasons. When intellect agrees with natural experience, intellectual understanding is valid, if not it is meaningless "[16]. It is clear that, in Patrizi's mind, practical intervention must be subordinate to science : in hydraulics, this means to knowledge on the nature of water. In his letter to the Duke on November 28, 1580, Patrizi boasts of being the first person to establish a general, orderly, real science of water resulting from natural and mathematical principles. He based his hydraulic proposals on this science.

Patrizi was not long involved in questions regarding the Reno River ; his involvement only ran from 1579 to 1581. Most likely Patrizi replaced another technician, Silvio Belli, who died in 1580. Silvio Belli first came to Ferrara in

13. The report dated July, 5, 1580 and the letter on November, 28, 1580 can be found in *ASM*, Confini dello Stato, 34A. For the letter on December, 9, 1981, see Archivio di Stato di Ferrara (hereafter called *ASF*), Archivio Storico Comunale (hereafter called *ASC*), Serie Patrimoniale, 27, n° 10. All this writings are now available in print ; see A. Fiocca, " Francesco Patrizi e la questione del Reno nella seconda metà del Cinquecento ", *op. cit.*

14. On the philosophical treatment of the subject of the water by Patrizi, see : G. Piaia, " Il tema delle acque in F. Patrizi ", *Quaderni per la storia dell'Università di Padova*, 29 (1996), 127-146.

15. F. Patrizi da Cherso, " Risposta alla scirittura di D. Scipio di Castro sopra l'arenamento del Po di Ferrara ", in D. Aguzzi Barbagli (ed.), *Lettere e opuscoli inediti*, Firenze, 1975, 373.

16. F. Patrizi da Cherso, " Sopra l'arenamento del Po di Ferrara ", in D. Aguzzi Barbagli (ed.), *Lettere e opuscoli inediti*, Firenze, 1975, 389.

1573 when Duke Alfonso II decided to resolve the hydraulic problems and called in a professional man who was at that time *proto alle acque* for the Republic of Venice. Indeed in the 16th century, the Republic of Venice was leader in hydraulics. In Venice, several factors came together, setting the stage for the study and development of an adequate policy of public intervention : the importance of hydraulics for that city and its complexity due to contrasting interests of the lagoon and mainland. A new, permanent magistracy was set up, including three sort of technicians, the so-called *proti alle acque*, respectively delegated to resolve the problems of the lagoon, the shores and the rivers[17].

Silvio Belli was born in Vicenza in around 1510[18]. His father, Francesco, was an accountant for the Mont-de-Piété and his uncle, Valerio, was a renown engraver. In 1555, together with the architect Andrea Palladio, Belli helped found the Olympic Academy in Vicenza. The aim of this Academy was to develop mathematics and, in general, the positive sciences. In 1556 the Academy appointed Belli " reader of the Sphere and other mathematical subjects ". In 1556 he was also appointed soprastante or chief engineer for the municipality of Vicenza and he collaborated with Palladio in construction of the loggia of the Basilica in Vicenza.

In 1565, in Venice, he published a book on practical mathematics dealing with the measurement operations associated with land surveying. This book, *Libro del misurar con la vista*, also had subsequent editions during the 16th century. On December 19, 1566 he was appointed proto alle acque for the Venetian Republic. In 1573 he published his own revision of the Euclidean theory of proportions in his book, *Della proportione et proportionalità communi passioni del quanto*, which, together with the previously mentioned work, had a further posthumous edition in 1595.

In 1573 Alfonso II of Este called Belli to Ferrara, asking his opinion on the real possibility of renavigating the Po di Ferrara[19]. Belli returned to Ferrara for other commissions — in particular for judicial inspections of the canals for the great land reclamation of the Polesine of Ferrara initiated in 1564 by the Estense Family — and was finally appointed Ducal Engineer. His last report to the Duke is dated June 1579 and he died shortly thereafter[20].

---

17. On the hydraulics of the Venetian Republic in the16[th] century, see S. Escobar, " Il controllo delle acque : problemi tecnici e interessi economici ", in G. Micheli (ed.), *Storia d'Italia. Annali 3. Scienza e tecnica nella cultura e nella società dal Rinascimento ad oggi*, Torino, 1980, 85-153, and S. Ciriacono, " Scrittori d'idraulica e politica delle acque ", in G. Arnaldi, M. Pastore Strocchi (eds), *Storia della cultura veneta*, 3/II, Vicenza, 1980, 492-512.

18. For a biography of Belli, see F. Barbieri, " Belli Silvio ", *Dizionario Biografico degli Italiani* (hereafter called *DBI*), *ad vocem*. For Silvio Belli's scientific and professional activity see : A. Fioca, " Silvio Belli ingegnere : empiria e matematica nella cultura tecnica del Rinascimento ", in D. Biancardi and F. Cazzola (eds), *Acque e Terre di Cofine*, Editrice Cartografica, 2000, 15-49.

19. Copy of Belli's report to the Duke on August 23, 1573, containing his opinion on three different projects for the River Reno can be found in *ASF, ASC*, Serie Patrimoniale, 25, n° 12.

Negotiations with Bologna for a new Reno-line were resumed in 1577 when Pope Gregorio XIII appointed a special board on the question, headed by Cardinal Filippo Guastavillani and supported by Papal Engineer Scipio di Castro. Representatives from Bologna and Ferrara attended the meeting held in Rome that year[21]. The Representatives of Ferrara produced a document written by Belli maintaining that the inlet of the Reno into the Po River was what caused the riverbed to fill up with earth deposits. They also proposed a means to solve the problem : canalizing all seven streams flowing from the Reno to the Adriatic Sea *(i.e.* the Reno, Savana, Idice, Sillaro, Santerno, Senio and Lamone), passing them through the five remaining marshes of the old Padusa and letting them flow into the Adriatic Sea. In the meantime, while waiting for the new canal to be built, they proposed closing the existing mouth of the Reno into the Po, letting its waters flood the marshes south of Ferrara and opening a lower outlet for these waters into the Po di Primaro[22].

The main objection the Bologna representatives had to this proposal concerned the rise in the marsh water level that would be caused by inflow of the Reno waters. Belli's reply to this objection is an interesting mathematical calculation also worked up in the treatise : *Discorso di Silvio Belli sopra l'arrenamento del Pò con le provisioni da farsi levando principalmente il Reno et mettendolo nelle valli*, a copy of which can be found in the Estense Library in Modena[23]. To demonstrate that the fears expressed by the Bolognese were excessive, Belli calculated how long a single flood of the Reno would have to last to increase the marsh level by one foot. Considering the geometrical measurements of the mashes and the wet cross-section of the Reno River, and by considering a Reno flow rate of four miles per hour, Belli calculated that the flood would have to last nearly 24 hours. A flood, however, is usually much shorter and, in any case, Belli's project also called for an outlet from the marshes. For the above reasons Belli concluded that the main objection forwarded by the Bolognese was unfounded.

In my opinion, this example is quite meaningful as it shows that, by the 1570s, some practical hydraulics engineers had already identified new

20. On the 16th century great land reclamation of the Polesine of Ferrara and, in particular, on the problems which caused Bell's judicial inspections see : F. Cazzola, " La bonifica del Polesine di Ferrara dall'età estense al 1885 ", *La grande bonificazione ferrarese*, Ferrara, 1987, 103-251 (*cf.* 165-168).

21. Two letters by Belli concerning the negotiation written from Rome on March 13, 1577 and on April 16, 1577 can be found in *ASM*, Confini dello Stato, 33A, and print in A. Fioca, " Silvio Belli ingegnere : empiria e matematica nella cultura tecnica del Rinascimento ", *op. cit.*, 41, 48.

22. F. Patrizi da Cherso, *Lettere e opuscoli inediti, op. cit.*, 424-427. An incomplete manuscript of Belli's writing can be found in Biblioteca Estense, Modena (hereafter called *BEM*), Autografoteca Campori, Belli Silvio.

23. Ms. It. 349 = P.5.10. Two other hand-written documents on the objections forwarded by Bologna's representatives and Belli's reply can be found in *ASM*, Confini dello Stato, 34A.

approaches to the problems of water motion. In addition, this is the first attempt to subject river hydraulics to mathematics : *i.e.* geometry.

At the end of the 16th century, Alfonso II passed away and the Dukedom of Ferrara came under Papal rule ; the onset of a new age in the history of the Reno River. The architect-engineer Giambattista Aleotti (1546-1636) dominated the early part of the following century.

Aleotti is a very complex, composite figure, a technician who could be considered one of the last great engineers of the Renaissance. Besides hydraulics he was involved in many other fields : civil and military architecture, artillery, perspectives and mechanics associated with theatrical scenography and scenecraft, scientific instrumentation as he developed his own surveying measures, music composing a treatise on this subject[24].

Aleotti was involved in hydraulics from at least the early 1580s although his masterwork in this field dates from the turn of the century. In his tract *Dell'interrimento del Po di Ferrara e divergenza delle sue acque nel Po di Ficarolo*, dated 1601, Aleotti put Belli's mathematical invention to different use[25]. His aim was to prove that what had caused the Po River bed to fill up with earth was the slow down of the Reno's turbid waters just beyond its inlet into the wider bed of the Po, thus depositing the suspended soil particles. Like Belli, Aleotti used a geometrical model to represent the motion of the water. The model he used consisted of rectangular parallelepiped of water, the base of which was equal to the wet cross-section of the river, its third dimension or height being equal to the space covered by the water in a given amount of time. Aleotti calculated the slow-down of the Reno's waters by comparing the measurement of the heights of two parallelepipeda representing the same amount of water (*i.e.* the Reno flood) in the two riverbeds (*i.e.* the Reno and the Po). Taking the Reno flow velocity as 4 miles per hour, considering a flood lasting approximately 8 hours, and taking the geometrical measurement of the wet cross-sections for the Reno and Po rivers, respectively, to be 15 and 35 square perches, Aleotti calculated that the Reno water entering the Po River bed covered 13 miles in 8 hours ; that is a velocity of 1.6 miles an hour[26].

According to these examples, Belli and Aleotti were able to calculate the river flow rate using the velocity of its waters. Aleotti knew that the velocity of running water changes when there is a change in the cross-section. Moreover,

24. The University of Ferrara dedicated a meeting to G. Aleotti in December 1996 ; the proceedings are now available : A. Fiocca, " Giambattista Aleotti e la *scienza et arte delle acque* ", in A. Fiocca (ed.), *Giambattista Aleotti e gli ingegneri del Rinascimento*, Firenze, 1998.

25. G.B. Aleotti, *Dell'interrimento del Po di Ferrara*, published posthumous by L.N. Cittadella, Ferrara, 1847, 59-77.

26. Actually Aleotti does not use the word velocity, but he says that " the River Reno ran, as it runs, four miles an hour " (*Il Reno correa, siccome corre, quattro miglia per ora*). Also Belli does not use the word velocity, but he says that the Reno's flood moves so fast that it covers four miles an hour (*si muove così veloce che fa per ora quattro miglia*).

using his geometrical model, he was able to measure this variation in " velocity ".

Aleotti's master work in hydraulics is the *Hydrologia overo della scienza et arte del ben regolare l'acque*, whose autographic copy is housed in the Biblioteca Comunale Ariostea in Ferrara (Ms. Cl. I, 749), a vast treatise addressed to hydraulic engineers which, had it been published, would have been the first treatise of its sort[27]. In effect, *Hydrologia* is a summa of Renaissance hydraulic tradition, drawing particular theoretical inspiration from a chapter in the encyclopaedic work *De rerum varietate* published by Girolamo Cardano in Basel in 1557[28].

To see a new trend in hydraulics we have to wait for a work by one of Galileo's pupils, Benedetto Castelli, first published in Rome in 1628 under the title : *Della misura dell'acque correnti*. This work is considered a real turning point in hydraulics because it presents a radical methodological shift, transforming the practical problem of river control into a new branch of mathematical science. Castelli's main contribution to hydraulics is the well-known law of steady flow continuity, given in the initial proposition of his book. Demonstration is based on the principle of conservation of volumes, equalizing flow rates in all river cross-sections : " Sections of the same river discharge equal quantities of water in equal time, even if the sections themselves are unequal ", " Given two sections of a river, the ratio between the quantity of water which passes the first section and that passing the second is proportional to the ratio between the first and second sections and to that of the first and second velocities ".

According to Castelli, he undertook the study of water motion after being employed to accompany Monsignor Ottavio Corsini to visit the waters of the Legations of Bologna, Ferrara and Ravenna in 1625. At that time the most important problem faced by the three Legations was still the course of the Reno River which had been diverted in 1604, closing the mouth into the Po and sending its waters into the Sammartina, a marshy area south of Ferrara. This diversion was part of an ambitious land reclamation plan and was intended to be temporary to dig out the Po di Ferrara and Po di Primaro riverbeds so as to restore navigation. Actually the project came to naught while the Reno floods expanded into the plain between Ferrara and Bologna transforming the area into a large marsh. For nearly two centuries the Bolognesi continued their promotion of the project to lead the Reno back into the Po. However,

27. Aleotti's treatise has been recently published under the title : *Della scienza et dell'arte del ben regolare le acque*, in M. Rossi (ed.), Modena, 2000. Unfortunately, the choises by the editor are questionable : the main objection being thet he published a non-autographic copy, instead of Aleotti's autograph.

28. On Aleotti's theoretical ideas in hydraulic field see A. Fiocca, " Giambattista Aleotti e la *scienza et arte delle acque* ", *op. cit.*, 47-101, and in particular 56-68.

the conflict of interests between Ferrara and Bologna, and the opposition of neighboring States, caused the project to fall through.

In his work Castelli did not acknowledge any cultural debt to the practical men for his hydraulic conceptions ; on the contrary, he started by criticizing the technicians and the mistaken way they had measured running water considering only the river cross-section, neglecting velocity. Actually, according to my research results, in spite of Castelli's criticism, I feel his new, fundamental result clearly shows the rational development by a new scientific mind, of the store of knowledge accumulated from the hydraulic scholars of earlier generations.

THE LEANING TOWER EXPERIMENT IN CONTEXT : BORRO, BUONAMICI AND GALILEO ON FALLING BODIES

Michele CAMEROTA, Mario O. HELBING

" At this time, it appearing to him that for the investigation of natural effects there was necessarily required a true knowledge of the nature of motion, there being a philosophical and popular axiom that " Ignorance of motion is ignorance of nature ", he quite gave himself over to its study ; and then, to the great discomfort of all the philosophers, through experiences and sound demonstrations and arguments, a great many conclusions of Aristotle himself on the subject of motion were shown by him to be false which up to that time had been held as most clear and indubitable, as (among others) that speed of unequal weights of the same material, moving through the same medium, did not all preserve the ratio of their heaviness assigned to them by Aristotle, but rather, these all moved with equal speeds, he showing this by repeated experiments [*esperienze*] made from the height of the Leaning Tower of Pisa in the presence of other professors and all the students "[1].

These words from Vincenzo Viviani's Galilean biography refer to the years between 1589 and 1592, when Galileo held the chair of mathematics at the university of Pisa.

In telling the story of the famous, much questioned Leaning Tower experiment, Viviani's account supplies us with one of the few (surely the best known) pieces of information about the earliest stage of Galileo's scientific career.

Although the Leaning Tower experiment in itself has been very extensively debated and very carefully analyzed, less attention has been paid to the context in which it was allegedly performed.

---

1. V. Viviani, *Racconto istorico della vita di Galileo*, Engl. trans. in S. Drake, *Galileo at Work. His Scientific Biography*, Chicago and London, 1978, 19.

Thus, apart from a few noteworthy studies on the subject[2], we still lack a good grasp of the cultural milieu of Galileo's Pisan years, including the " other professors " mentioned by Viviani in his account of the Leaning Tower experiment.

Indeed, in the last decades of the sixteenth century, Galileo was not the only natural philosopher to perform an experiment of fall, dropping bodies from a high place.

From the studies of Raffaello Caverni we know that the famous and very much disputed Leaning Tower experiment had several predecessors[3]. As noticed by Thomas Settle : " by the 1540s a few, even if not yet a great many, natural philosophers had already tested empirically what they took to be Aristotle's assumption about natural motion and found them wanting "[4].

Among these early experimenters we can find also the philosophy professors at the university of Pisa, Girolamo Borro (1512-1592)[5] and Francesco Buonamici (1533-1603)[6], whose intellectual rivalry gave rise to a long-lasting quarrel on the problem of the " motion of light and heavy bodies " (*quaestio de motu gravium et levium*)[7]. Galileo's early writings on dynamics (*De motu antiquiora*)[8] were largely connected to this debate[9].

2. See especially : N. Badaloni, " Il periodo pisano nella formazione del pensiero di Galileo ", *Estratto da " Saggi su Galileo Galilei "*, Florence, 1965 ; E. Garin, " Galileo e la cultura del suo tempo ", in E. Garin (ed.), *Scienza e vita civile nel Rinascimento italiano*, Bari, 1965 (repr. 1993), 109-146 ; C.B. Schmitt, " The Faculty of Arts at Pisa at the Time of Galileo ", *Physis*, XIV (1972), 243-272 ; C.B. Schmitt, " The University of Pisa in the Renaissance ", *History of Education*, III (1974), 3-17 ; C.B. Schmitt, " The Studio Pisano in the European Cultural Context of the Sixteenth Century ", *Firenze e la Toscana dei Medici*, I, Florence, 1983, 19-36 ; P. Zambelli, " Scienza, filosofia e religione nella Toscana di Cosimo I ", in S. Bertelli, N. Rubinstein and C.H. Smith (eds), *Florence and Venice : Comparisons and Relations*, II, Florence, 1980, 3-52. For an outline of the teaching of philosophy and mathematics in Pisa, see M. Ioffrida, " La filosofia e la medicina (1543-1737) ", *Storia dell' Università di Pisa. 1343-1737*, I, Pisa, 1993, 289-338 ; C. Maccagni, " La matematica ", *Storia dell' Università di Pisa. 1343-1737*, I, 339-362.

3. See R. Caverni, *Storia del metodo sperimentale in Italia*, IV, Florence, 1891-1900, 266-276.

4. T.B. Settle, " Galileo and Early Experimentation ", in R. Aris, T.H. Davis and R. Stuewer (eds), *Springs of Scientific Creativity. Essays on Founders of Modern Science*, Minneapolis, 1983, 3-20, 16.

5. On Borro see : N.W. Gilbert, *Renaissance Concepts of Method*, New York, London, 1960, 186-195 ; E. Garin, *Galileo e la cultura del suo tempo*, 123-126, 141-142 ; G. Stabile, Borri Girolamo, *Dizionario biografico degli italiani*, XIII (1971), 13-17 ; C.B. Schmitt, " Girolamo Borro's Multae sunt nostrarum ignorationum causae (Ms. Vat. Ross. 1009), in E.P. Mahoney (ed.), *Philosophy and Humanism. Renaissance Essays in Honor of Paul Oskar Kristeller*, Leiden, 1976, 462-476 ; C.B. Schmitt, " Borro Girolamo ", *Dictionary of Scientific Biographies*, XV (1978), 44-46 ; L. Conti, " Girolamo Borro : cardiocentrismo e *perfettione delle donne* ", in L. Conti (ed.), *Medicina e biologia nella rivoluzione scientifica*, Assisi, 1990, 65-106 ; A. De Pace, " Galileo lettore di Girolamo Borri nel *De motu* ", *De motu. Studi di storia del pensiero su Galileo, Hegel, Huygens e Gilbert, Quaderni di Acme*, 12 (1990), 3-69 ; M. Ioffrida, *La filosofia e la medicina*, I, 296-301.

6. On Buonamici see M.O. Helbing, *La filosofia di F. Buonamici, professore di Galileo a Pisa*, Pisa, 1989, and the bibliographical references listed there, p. 41 n° 1.

7. See M.O. Helbing, *La filosofia di F. Buonamici, professore di Galileo a Pisa*, 54-57.

8. See G. Galilei, *Opere*, I, a cura di A. Favaro, Florence, Edizione Nazionale, 1890-1909, 243-419 (repr. 1968), [hereafter referred to as *Opere*].

Girolamo Borro's experiment of fall has already been discussed by Galilean scholars, especially by Charles Schmitt and Thomas Settle[10]. They have carefully analyzed the experimental report of the philosopher from Arezzo, without, however, taking into due account the theoretical context of Borro's argument.

In a chapter of his book *De motu gravium and levium* (published in Florence in 1575), Borro maintains that he performed an experiment projecting pieces of wood and lead from a high window. The question under discussion concerned the texts 29 and 30 of the fourth book of Aristotle's *De caelo*[11], which had different interpretations by Themistius and by Averroes : the former stated that air did not weigh in its natural place, whereas the latter agreed with Aristotle's opinion that " in its own place every body has weight except fire, even air "[12].

On the basis of these different positions, Themistius and Averroes gave conflicting interpretations of another Aristotelian statement, which claimed that : " in air [...] a talent of wood is heavier than a mina of lead, but in water is lighter "[13]. In the Aristotelian tradition, wood was regarded to be a mixed body (containing all the four elements) with a prevalence of air, whereas lead was considered to be mostly compounded of water and earth. In conformity with his opinion that air did not have weight in its own place, Themistius maintained that wood was heavier in air since it consisted of a greater part of water and earth than lead. On the other hand, Averroes agreed that wood was heavier in air, but he held that it was heavier just because of the weight of air in its own place[14].

In order to confirm Averroes' theory that Borro tells us that he threw from the window of his house pieces of lead and wood of about the same weight (he claims that he did not deem it necessary to weigh them with a balance). According to Borro, if Themistius were right the air contained in the piece of wood would not have weight in air, and so it would not be able to push the body downward. Then, a piece of wood and a piece of lead of equal weight should touch the ground at the same time. On the other hand, if Averroes' opinion is correct, the wood should fall faster than the lead, since the air, having weight in its own place, would be able to carry it downward : " On this

9. See M. Camerota and M.O. Helbing, " Galileo and the Pisan Aristotelilanism : Galileo's *De motu antiquiora* and the *Quaestiones de motu elementorum* of the Pisan professors ", *Early Sciences and Medicine*, V, n° 4, 2000, 319-365.

10. See C.B. Schmitt, *The Faculty of Arts at Pisa*, 269-271 ; T.B. Settle, *Galileo and Early Experimentation*, 6-8.

11. *Aristotelis Opera cum Averrois Commentariis*, Venetiis, Apud Iuntas, 1562-1574 [hereafter abbreviated as Aristotelis *Opera*], V, ff. 256v-257r. See *De caelo*, 311b 1-13.

12. *De caelo*, 311b 8-9 ; Aristotle, *On the Heavens*, Engl. trans. by W.K.C. Guthrie, Cambridge Mass., 1939, 355 (repr. 1986).

13. *De caelo*, 311b 3-4. A " mina " is equivalent to 60 talents.

14. See Averroes on *De caelo*, in Aristotelis *Opera*, V, f. 257v.

question, — writes Borro — one does not have to listen to Themistius, who holds that a great piece of wood is heavier than a small piece of lead not because of air (which he denies to have weight in its own place), but because of the greater quantity of water and earth contained in it. For this reason, Averroes, in the fourth book of *De Caelo*, on commentary 31[15], strongly attacked Themistius, especially giving the following argument. If in the wood and in the lead there were the same force coming from water and earth, then — as Themistius wants — two pieces of wood and lead of the same weight would fall through air with the same speed. Yet, in accordance with natural truth, their weights are different. In fact, there are three heavy elements in wood : air, water and earth, all of which tend to the center, and so they draw downward the bodies in which they are contained. On the other hand, there is a smaller quantity of air in the lead, and thus in it there will be less weight coming from the aerial element. Thus, taking portions of the same weight (for instance, one talent), wood will descend faster than lead in air. And this was precisely what was experienced by some of those who came to listen to us a few days ago "[16].

Borro tells that, during a meeting in his house, the averroistic position was questioned at length by some of his guests, who held different opinions. Thus, in order to find a solution to the question, Borro and his friends decided to perform an experiment : " [...] and since, among us, discussions always grew without coming to a conclusion, we resorted, as to a holy anchor, to experience, the teacher of all things. We took, then, pieces of wood and lead of the same weight. This, at least, as far as one could guess from the appearances, because we did not think necessary to weigh them with a balance, believing that — as for the experiment we were to perform — it was enough to establish their weight by sight. Therefore, after having found two pieces of equal weight, we threw them from a rather high window of our house at the same time and with the same force. The lead descended more slowly, being above the wood, which fell to the ground first. All of us were waiting for the result of this occurrence, and we all saw the latter [*i.e.* the wood] fall headlong. And not only once, but many times we tried the test with the same results. Compelled by the results of the experiment, all adhered to our opinion. Therefore, either by reason, by experiment, and by authority, it is appropriate to conclude that air has some weight in its own place, so that a piece of wood, which contains a greater part of air than a piece of lead of equal weight, descends more swiftly in air "[17].

Hence, Girolamo Borro invokes experience to disprove Themistius' opinion, and to support Averroes' theory that air has weight in its own place. On the

---

15. Borro's reference is mistaken : Averroes criticizes Themistius in commentary 30.
16. G. Borro, *De motu gravium et levium*, Florentiae, G. Marescottius, 1575, 214.
17. G. Borro, *De motu gravium et levium, op. cit.*, 215.

other hand, in Buonamici's treatise *De motu* (issued in 1591) we find a defense of Themistius' view.

After a *resumé* of the *quaestio*[18], Buonamici refers to " somebody " (*quisquam*) who believes that the swifter fall of wood in air proves the correctness of Averroes' theory[19]. At that time, it was usual for philosophers not to mention by name their adversaries. It is therefore likely that by the pronoun " somebody ", Buonamici was implicitly referring to his colleague and rival Borro, who — as we have seen — considered the alleged swifter fall of wood as strong evidence for the Averroistic claim that air has weight in its place. This assumption seems to find further confirmation in Buonamici's next argumentation, whose development is to a large extent *ad hominem*.

We have seen that Borro did not weigh the pieces of wood and lead he used for the experiment. This circumstance is noted by Buonamici, who remarks that it is necessary to weigh the two bodies with a balance before throwing them from a high place. Furthermore, Buonamici maintains that, in order to find which of the two bodies fell more swiftly, one had to look at the impact (*ictus*) on the ground : " Nor does sense experience urge us to believe the contrary [of Themistius' opinion]. In fact, you can compare a body mostly compounded of air and one mostly compounded of earth by throwing them from a great height, only if the bodies have an equal weight when hanging on to a balance. And [then] you can experience which of them causes the greater impact "[20].

Finally, after a long discussion, Buonamici states that one has to distinguish between two different notions of weight : the *gravitas extensive* (corresponding to absolute weight) and the *gravitas secundum gradum* or *gravitas intensive* (by and large, similar to the concept of specific gravity). Furthermore, he argues that the natural motion seems more closely related to the *gravitas intensive* than to the *gravitas extensive*[21]. Buonamici's statement here is rather obscure and ambiguous, in view of the fact that in several other places in his book, he expresses his strict adherence to the traditional Aristotelian position, contending that the speeds of different bodies falling in the same medium are proportional to their absolute weights. Anyway, Buonamici's final verdict on Borro's argument is very clear : " Then, with the help of this distinction

---

18. " Averroes criticized Themistius in this way: a body is heavier than another when it has a larger number of gravities [*gravitates*]. But in wood there are three gravities [*gravitates*], namely that of air, that of water, and that of earth, whereas in lead there are just two : that of water and that of earth. So wood is heavier in air than lead ". F. Buonamici, *De motu Libri X*, Florentiae, Apud B. Semartellium, 1591, 482.

19. " Nor should somebody, from the occurrence that wood descends more swiftly in air than stone or lead, being in it [wood] prevalent the parts of air, attribute the cause of wood's speed to air, on the grounds that air would move in its own place more swiftly than the other bodies ". F. Buonamici, *De motu Libri X, op. cit.*, 483.

20. F. Buonamici, *De motu Libri X, op. cit.*, 483.

21. See F. Buonamici, *De motu Libri X, op. cit.*, 484-485.

[between the two notions of gravity], Themistius will retort the statement [*i.e.* the theory of Averroes and Borro], claiming that the body which has more " intensive gravity " (*gravitas secundum gradum, i.e. gravitas intensive*), moves more swiftly. And so the argument [of Borro] should be considered false, because in wood there is not a more intense gravity than in lead. Still, even if we take into account the absolute weights (*extensive gravitates*), the argument should not be granted, since reasons and experience, as well as the assumptions of Aristotelian philosophy demand this verdict "[22].

Although Buonamici does not claim explicitly to have performed experiments, he also calls upon experience. He seems to be quite sure that, once observed the procedure of weighing the bodies before throwing them, Borro's argument will be disproved. At the same time, he attempts to develop a theoretical counterargument by taking into account specific weights. Even though his account is rather confused and puzzling, it can be safely concluded that Buonamici was engaged in a dispute in which empirical findings were placed in the foreground as a primary source of evidence.

Let us turn now to Galileo. In Chapter 22 of the Treatise *De motu*, we can find a discussion of the same problem treated by Buonamici and Borro. The latter is mentioned expressly in the margin of the page, even though in the text Galileo refers generically to " Averroes and his followers " (*Averroes et qui eum secuntur*) : " They hold that air is heavy in its own region, from which it follows that things which have more air are heavier in the region of air (and this is also Aristotle's opinion). Thus [they say] a wooden sphere, for example, since it has more air in it than a leaden one, has three heavy elements, air, water, and earth ; while the leaden one, since it has less air in it, has, as it were, only two heavy elements : the result of this is that the wooden sphere falls [in air] more swiftly than the leaden "[23].

Galileo raised several objections to the Averroistic position. One of them seems to parallel Buonamici's conclusion : " if the velocity of the [natural] motion of a body depends on its weight, as everyone holds, and if the leaden sphere has earth and water in place of the portions of air that are in the wooden sphere, and if earth and water are heavier than air, as we can readily believe, then will not the lead be heavier and fall more swiftly ? "[24].

Moreover, immediately after this argument, Galileo points out a new comparison with iron : " And as for what they say about iron and lead, to show that air adds to the weight, if lead is heavier because it has more air, then wood will be heavier than both iron and lead, since it has more air than either of them "[25].

22. F. Buonamici, *De motu Libri X, op. cit.*, 485.

23. *Opere*, I, 333 ; Engl. trans. in I.E. Drabkin and S. Drake, *Galileo Galilei : " On Motion "* and *" On Mechanics "*, Madison, 1960, 106 [hereafter abbreviated as *On Motion*].

24. *Opere*, I, 334 ; *On Motion*, 106.

25. *Opere*, I, 334 ; *On Motion*, 106-107.

Galileo is referring here to another point in Borro's argumentation : according to the Aristotelian philosopher, lead would be heavier in air than iron, because the former contains a greater part of air than the latter. The question was discussed by Buonamici, too. A comparison between the statements with which Buonamici, Borro and Galileo introduced the question can show how much Galileo's text is linked to the Pisan disputation :

Borro

And lead too, being heavier than iron, is heavier in air than iron as a consequence of the parts of air, which are more in rare lead.

Buonamici

Here, it is raised against us the objection of lead, which is rarer than iron, but still heavier than it.

Galileo

And not content with this, they also say that rare lead is heavier than dense iron in air for the reason that there are more parts of air in rare lead than in dense iron[26].

Both Buonamici and Galileo refused to admit that there are more parts of air in lead than in iron. Buonamici held that lead was denser, whereas iron contained air in its pores[27]. Galileo, on the other hand, tackled the subject by making the above mentioned comparison between wood, lead and iron, and claiming that, if one assumes that something is heavier in air because it has more air, then wood would be heavier than lead or iron. Therefore : " if the great quantity of air which is in wood makes the wood move more swiftly, then it will always move more swiftly, so long as it is in the air "[28].

This would be the case if Borro were right, but, invoking again experience, Galileo claimed that it : " shows us the opposite "[29].

At this point, one could expect that Galileo — like Buonamici — would deny that wood descended more swiftly than lead or iron. Yet, Galileo's experimental report is quite different from both Buonamici's and Borro's : " it is true that wood moves more swiftly than lead in the beginning of its motion ; but a little later the motion of the lead is so accelerated that it leaves the wood behind it. And if they are both let fall from a high tower, the lead moves far out in front. This is something I have often tested "[30].

26. G. Borro, *De motu gravium et levium*, 232 ; F. Buonamici, *De motu*, 485 ; *Opere*, I, 333.

27. *Cum caeteroqui ferrum inaequaliter mixtum sit, et in poris aliquam aeris portionem contineat, itaque ferrum in aere levius, credo etiam in aquis.* F. Buonamici, *De motu, op. cit.*, 485.

28. *Opere*, I, 334 ; *On Motion*, 106-107.

29. *Opere*, I, 334 ; *On Motion*, 107.

30. *Opere*, I, 334 ; *On Motion*, 107.

As showed by Thomas Settle, the effect of seeing the light bodies preceding the heavy ones in the first part of their fall is not an optical illusion, but a real phenomenon[31]. We cannot go into details here, and so we will not take into account all the experimental tests alleged by Galileo in *De motu antiquiora*[32]. We limit ourselves to remarking that Galileo used the above mentioned observation to emphasize the shortcomings and the explanatory inadequacy of Aristotelian dynamics (as represented in the debate between Borro and Buonamici), coming to the conclusion that : " we must try to derive a sounder explanation on the basis of sounder hypotheses "[33].

Galileo's " sounder explanation " was based on the Hipparchean theory of acceleration, which accounted for the accelerated fall of bodies by invoking the gradual decrease of the " lightness " (*virtus impressa*) and the concomitant increase of the effects of weight[34]. This explanation had been disseminated in the Renaissance through the translation of Simplicius' commentary on *De caelo*[35]. As remarked by Adriano Carugo, Galileo probably became acquainted with it reading Benito Pereira's *De communibus omnium rerum naturalium principiis*[36]. It is, however, noteworthy that Galileo introduced this argument in the same theoretical context as that figuring in the dispute among Pisan professors, who, on their part, also discussed the Hipparchean theory[37].

So, we believe that, as regards the experimental texts on falling bodies, Galileo's statements in *De motu antiquiora* must be read in close connection to the Pisan debate on the " motion of the elements " (*de motu elementorum*). Furthermore, it came out that the treatment of the subject was heavily grounded on experimental considerations, since all the participants involved in the debate (Borro, Buonamici and Galileo) used empirical evidence to dismiss the opinions of their adversaries.

A further clue of this experimental trend is provided by a later document — Giorgio Coresio's *Operetta intorno al galleggiare de' corpi solidi*[38] — dating 1612. At that time, Galileo was engaged in a heated dispute on hydrostatics with several Aristotelian philosophers. Coresio's booklet was just one of the

31. The most probable explanation is related to the increased muscular tension required to hold the heavier body : this could produce an imperceptible delay in its release. See T.B. Settle, *Galileo and Early Experimentation, op. cit.*, 12-14.

32. For the test performed from " a high tower " (*alta turris*), see *Opere*, I, 263, 273, 317, 326, 329, 334, 406, 407.

33. *Opere*, I, 334 ; *On Motion*, 107.

34. See *Opere*, I, 315-323.

35. Simplicius, *Commentaria in quatuor libros Aristotelis De coelo*, Venetiis, Apud Haer. I. Scoti, 1584, ff. 77-78.

36. See A. Carugo, " Les Jésuites et la philosophie naturelle de Galilée : Benedictus Pererius et le *De motu gravium* de Galilée, *History and Technology*, 4 (1987), 321-333.

37. See G. Borro, *De motu gravium et levium, op. cit.*, 239 ; F. Buonamici, *De motu, op. cit.*, 411.

38. See *Opere*, IV, 199-244.

several treatises produced by the Peripatetic group opposing Galileo. In the last pages of his work, Coresio attacked Jacopo Mazzoni, a friend of Galileo who held the chair of philosophy at Pisa from 1588 to 1598[39]. Coresio maintained that Mazzoni was wrong in denying that, in the case of bodies of the same material : " the whole moves more swiftly than the part "[40]. Mazzoni's mistake was that : " perhaps, he performed the experiment from his window, and, because the window was low, all the heavy substances went down from it evenly. But we did the test from the top of the cathedral tower of Pisa [*i.e.* from the Leaning Tower], confirming by experience the truth of Aristotle's statement that the whole of the same material and with a shape proportionate to [that of] the part, descended more quickly than it [*i.e.* the part] "[41].

This passage has already been noticed by Lane Cooper, who, however, concluded from it that Galileo did not perform the Leaning Tower experiment[42]. What is interesting for our purposes is that Coresio's discussion is a confirmation of the continuance of a controversy in Pisa, in which even Mazzoni (who held the chair in Pisa from 1588) took part[43].

At the same time, Coresio's text supplies us with a further evidence of the experimental trend that marked the Pisan dispute, whose the protagonists (Borro, Buonamici, Galileo and, very likely, Mazzoni) resorted to experience and to experimental tests as a way to settle the controversy, dismissing the arguments of adversaries.

From this point of view, Viviani's account of the Leaning Tower experiment is perhaps more reliable than contemporary historians of science are willing to acknowledge.

39. J. Mazzoni (1548-1598) was (along with F. Patrizi) the most representative figure of late Renaissance Platonism. On him see : F. Purnell, " Jacopo Mazzoni and Galileo ", *Physis*, XIV, 273-294 ; A. De Pace, " Archimede nella discussione su aristotelismo e platonismo di Jacopo Mazzoni ", in C. Dollo (ed.), *Archimede. Mito, tradizione, scienza*, Florence, 165-197.

40. *Opere*, IV, 242. Also Galileo held that : " in the case of bodies of the same material, the part and the whole move with the same speed ", *Opere*, I, 267 ; *On Motion*, 31.

41. *Opere*, IV, 242.

42. See L. Cooper, *Aristotle, Galileo and the Tower of Pisa*, Ithaca, New York, 1935, 28-29.

43. Notably, the issue of experimental conditions in performing tests of fall was widely discussed also in the hydrostatical controversy. See Camerota-Helbing, *Galileo and the Pisan Aristotelianism*, § 2.

CONTRIBUTORS

Aldo BRIGAGLIA
Università di Palermo
Palermo (Italy)

Michele CAMEROTA
Università di Cagliari
Cagliari (Italy)

Peter DAMEROW
Max-Planck-Insitut
Berlin (Germany)

Alessandra FIOCCA
Università di Ferrara
Ferrara (Italy)

Raymond FREDETTE
Université du Québec
Montréal (Canada)

A.C. GARIBALDI
Università di Genova
Genova (Italy)

Romano GATTO
Università della Basilicata
Potenza (Italy)

Enrico GIUSTI
Università di Firenze
Firenze (Italy)

M.O. HELBING
ETHZ, Cattedra di lingua
e letteratura italiana
Zürich (Switzerland)

Rosario MOSCHEO
Istituto di filosofia «Galvano
della Volpe»
Messina (Italy)

Pier Daniele NAPOLITANI
Università di Pisa
Pisa (Italy)

Pierre PINEL
INSA LESLA
Toulouse (France)

Jürgen RENN
Max-Planck-Insitut
Berlin (Germany)

Simone RIEGER
Max-Planck-Insitut
Berlin (Germany)

Ken SAITO
Osaka Prefecture University
Sakai, Osaka (Japan)

T.B. SETTLE
Istituto e museo di storia della scienza
Firenze (Italy)

Pierre SOUFFRIN
Observatoire de la Côte d'Azur
Nice (France)

Jean-Pierre SUTTO
Revel (France)

Abdel-Kaddous TAHA
INSA LESLA
Toulouse (France)

Roberta TASSORA
La Spezia (Italy)